To victims and disaster fighters of the Chernobyl and Fukushima-1 nuclear accidents.

# JOURNEY INTO THE FUTURE—STOP #1

Sergey Shestakov

seltmann+söhne

Everything is in our hands ...

Chernobyl … The name of the place has long since become a symbol of the enormous human tragedy. But the way our memory works means that in time details become less visible, reactions go numb, and that what happened becomes blurred, as if someone invisible reduces depth resolution in the lens of history. Chernobyl, that has shocked the entire world, is no exception. Quarter of a century passed since the night of the accident at the Chernobyl nuclear plant – April 25/26 1986, and we started to think that it was a one-time occurrence caused by improbable coincidence of many factors. Yes, it was horrible, but it will not happen again in human history. Everything changed on March 14 2011, when a huge 10-meter-high wave, caused by a very strong earthquake close to the shores of Japan, damaged the nuclear plant Fukushima-1.

# PREFACE

by Sergey Shestakov

When I planned my trip to the 30-km Exclusion zone last October, I did not think that photos I will make there will form a separate exhibition, let alone a book. The session was supposed to be just a start of a big project, a kind of meditation on the fate of the human race, which I plan to complete in 2012, and in was pure chance that Pripyat was chosen to be the first location. Everything I saw there affected me profoundly. We (me and my guide) walked alone in the completely empty city, and all the time I expected to hear in the next moment the noise of a passing car, or laughter of children playing in the yard, or to see a young mother with a pram… But it was the complete silence around us, the silence that was not disturbed even by birds. As we moved around the city and entered various buildings, the sense that all that is completely unreal, grew stronger. It was unreal not because that it can not be so, but because it SHOULD NOT be so!
There should not be empty cities and villages, the nature around us should not be destroyed, and people should not engage in self-extinction!

A month later I came back to Chernobyl, and decided to stay there overnight to try to gain more understanding of the pain of the people that have lived here and left their home. Probably, that experience charged the photographs with emotional intensity. And after the events at the Japanese nuclear plant, and all these parallels and comparisons with Soviet Chernobyl, I decided to make a separate exhibition, after all, and to publish this book. In these days, when all the world is absorbed with the tragedy of the Land of the Rising Sun, I want us all to think hard about where are we all going to.

The idea for the title came to me when I saw, in one of the Pripyat kindergartens, a half-rotten book by Zoya Voznesenskaya, "Journey to the Future". I only added the subtitle, "Stop No 1", because Chernobyl is not the terminal station. Today, when we are already at the stop No 2, people must decide, in what direction, by which line of the 'life metro' they will go further.

Everything is in our hands …

Obsession with 'future' is one of the 20th century's delusions and curses that were inherited from the 19th-century's optimism and progressism. But it is still not the reason to poison the 21st century by skepticism inherited from the 20th century. Salvation and danger come from the same source.

Mikhail Epstein. The future after the 'future'
From the 'To the New Model of the Future' series. Article 1[1]

# ART OF THE FUTURE. CHERNOBYL PHOTOGRAPHS BY SERGEY SHESTAKOV

by Wakana Kono

In calling our times 'the future after the future', Mikhail Epshtein refers to the breakdown of the USSR, i.e. the end of Communism, which was, according to him, one of the forms of 'futuromania' and 'futurocratia'[2]. In fact, the end of Utopian and anti-utopian projects and ideas, including the loss of faith in 'the peaceful atom', is a global trend of today's world.

In they search of models of the future 'after the future'. Contemporary artists present different images of the world. For example, Konstantin Batynkov, in his Running on the Waves series (the name refers to the story be Alexander Green) shows infinite catastrophes of the world 'after the future', and in every drawing, on the background of these natural and war disasters, there is an outline of a girl running on the waves. Leonid Tishkov demonstrates the possibility of a new utopian world in his images of 'Dubloids' and 'Divers', as well in the ideal space of the 'Ladomir', inspired by Velimir Khlebnikov. Stanislav Shuripa, an artist of the young generation that was growing up in the age of the computer, represents the virtual space in oils, creating a new artistic language for the depiction of the new world. And Alexander Ponomarev and Sergey Shestakov recently opened the 'Complementary Poles' exhibition, where they have shown the glaring world of the Earth's poles. The exhibition should act as a preview of an unprecedented event: the International Contemporary Art Biennale in Antarctica. According to Shestakov, "the festival that will take place in Antarctic will unite contemporary artists from different countries and will give them a possibility to make art together... It is very symbolic. Antarctic is, perhaps, the only territory on Earth where there are no barriers. Contemporary art has no barriers, too. It transcends any limits and stereotypes, and forms new meanings and energies". This project should become a stimulus for the future art in the world "without borders".

Shestakov's Journey to the Future. Stop No1 presents landscapes of the future 'after the future' in a different way. This series consists of photographs made by Shestakov in 2010 inside the 30-km Exclusion zone around Chernobyl. The idea of the series' name came to the artist when, in one of the abandoned kindergartens of Pripyat, he saw on a shelf the book by Zoya Voskresenskaya, 'Journey to the future'. The name captures the exact essence of the series: photographs of the former town's landscapes reflect the past, the present, and the future. In Shestakov's Chernobyl images, the perceptive reader is able to see today's situation at the Fukushima-1 nuclear plant, and a possible fate of the human race.

By creating works about the past, Shestakov speaks about the critical event in the contemporary world, and about the fate of the man, – and in this he is similar to Ilya Kabakov with his 'total installation'. Kabakov's total installations that represent the 'Soviet space' (10 Characters, Communal Kitchen, etc.), show not the Soviet world only, but, according to Kabakov, the global situation of the human race. The views of Pripyat houses and streets, filled with abandoned things – dirty dishes, yellowed exercise books, dusty toys, – resemble, to a certain extent, installations by Kabakov, in particular – The Toilet, and School No 6. Among the multitude of Chernobyl photographs by various authors, Schestakov's works stand out with their emotional intensity, and it is the quality that unites his photographs to installations by Kabakov.

In 1990, Kabakov said in a dialogue with Boris Groys: "… I feel as all the world and myself are filled with innumerable sounds: crackle, cackle, muttering, sounds of doors slammed, noise of cars – all of this is the noise of being. And, among all of the rest, – voices of other people around me and inside me. And all the time I feel the desire to answer to these voices, to announce my existence"[3], "the moment I see something, it immediately starts to sound inside me. The sound is reborn out a dead thing"[4]. And in Shestakov's photographs of Chernobyl we can hear voices of the former inhabitants: happy laugh of children, voices of teachers reading aloud, shouts of athletes. There are sounds of everyday life: a mother prepares dinner, a neighbor opens a door, the spring wind blows… The town itself is completely silent, but Shestakov's eyes and creative emotions give his photographs sounds of life and reflections of the people who are gone. And we too, through these photographs, share something of the author's emotions and experiences.

One could draw one more parallel, with works of the Japanese artist Kenji Yanobe, who, since the early 1990s, works in media art, photography, creates installations and performances on the subject of the radiation danger. Yanobe was born in 1965, in Osaka, the city that hosted the 1970 Expo. As a boy, he often played among the ruins of the Expo, and later that experience became for the young artist a metaphor of the end of the world and the fate of the human race, so Yanobe started to create works about "How man can survive the end of the world". In 1994, Yanobe visited Pripyat and made photographs of landscapes and his performances in the empty town – inside buildings, in the park near the ferris wheel. Visit to Chernobyl has strongly affected Yanobe's vision of the world. Having seen the real 'end of the world' and spoken with remaining residents of Chernobyl, Yanobe changed his main subject: instead of 'survival' it became 'revival'. The image of sun on the wall of one kindergarten in Pripyat gave new impetus for his work dedicated to revival (resurrection?) and bright future of children.

Evidently, it is not by accident that both Yanobe and Shestakov found the symbols for their Chernobyl series in an abandoned kindergarten: Yanobe – in the image of sun, and Shestakov in the name of the book, "Journey to the Future". To this name, Shestakov added the words "Stop No 1", "because Chernobyl is not the terminal station": this year, Fukushima became the tragic stop No 2. But the photographs by Sergey Shestakov, filled with emotions, breaths, voices – his own and of the gone inhabitants of the town, still give us hope to a journey to the sunny future.

1 http://veer.info/01/v1_bud.html
2 Ibid.
3 Ilya Kabakov, Boris Groys. Dialogues.
Vologda: Biblioteka Moskovskogo Kontheptualisma Germana Titova, 2010. P. 18.
4 Ibid. P. 19.

I thought I knew all about Chernobyl …

# INTRODUCTION
by Olga Sviblova

Most of the projects a curator and museum director works with are the result of a long and complex reflection. But, one way or another, on the bottom of any project is an emotional experience. I came across Chernobyl photographs by Sergey Shestakov quite recently and almost by chance. And I still remember the state of an emotional and existential shock I experienced at that moment.

I thought I knew all about Chernobyl, or, at least, very much. Chernobyl became a tragedy for many of my friends and could become a tragedy for members of my family. A few days before the nuclear accident, a group of Russian and Ukrainian poets was preparing a poetry night at the Chernobyl Nuclear Power Plant. In letters from Kiev to Moscow they joked: "everything will be all right, if nothing explodes". The radioactive explosion did touch a part of the poetry mission. Some of them later had ill children, while others have lost the ability to have children altogether.



For a quarter of century I read everything that there was to read on Chernobyl, watched films, photo- and video-reports. It was evidence that prevented forgetting, that kept reminding of the tragedy that has happened there. Sergey Shestakov's works do not so much refer to the past as they provoke feelings and thoughts about the future.

Every shot by Sergey Shestakov is a piercing metaphor, created from small and even tiny details. The whole project turns out to be a profound artistic generalization that especially 'rings' today, after the tragedy in Japan. This new situation changes the very sound of the name of the series, "Journey into the Future", derived from the name of a children book frozen in the Exclusion zone.

Sergey Shestakov's project is not as much about how difficult it is for people to calculate the possible danger of nuclear power plants, existing and those currently under construction, as it is about the actual measure of human responsibility for the achievements of civilization in general. One way, or another, all technologies are bound to get out of our control.

Artists of all times were attracted to the subject of the Apocalypse. "Journey into the Future by Sergey Shestakov is one more work of this kind. Today, in the beginning of the 21st century, which is extremely focused on modernization, innovations, new technologies, could be interpreted as a warning.

In his project on Chernobyl, Sergey Shestakov managed to achieve the ultimate articulation of the artistic message.

# WORKS

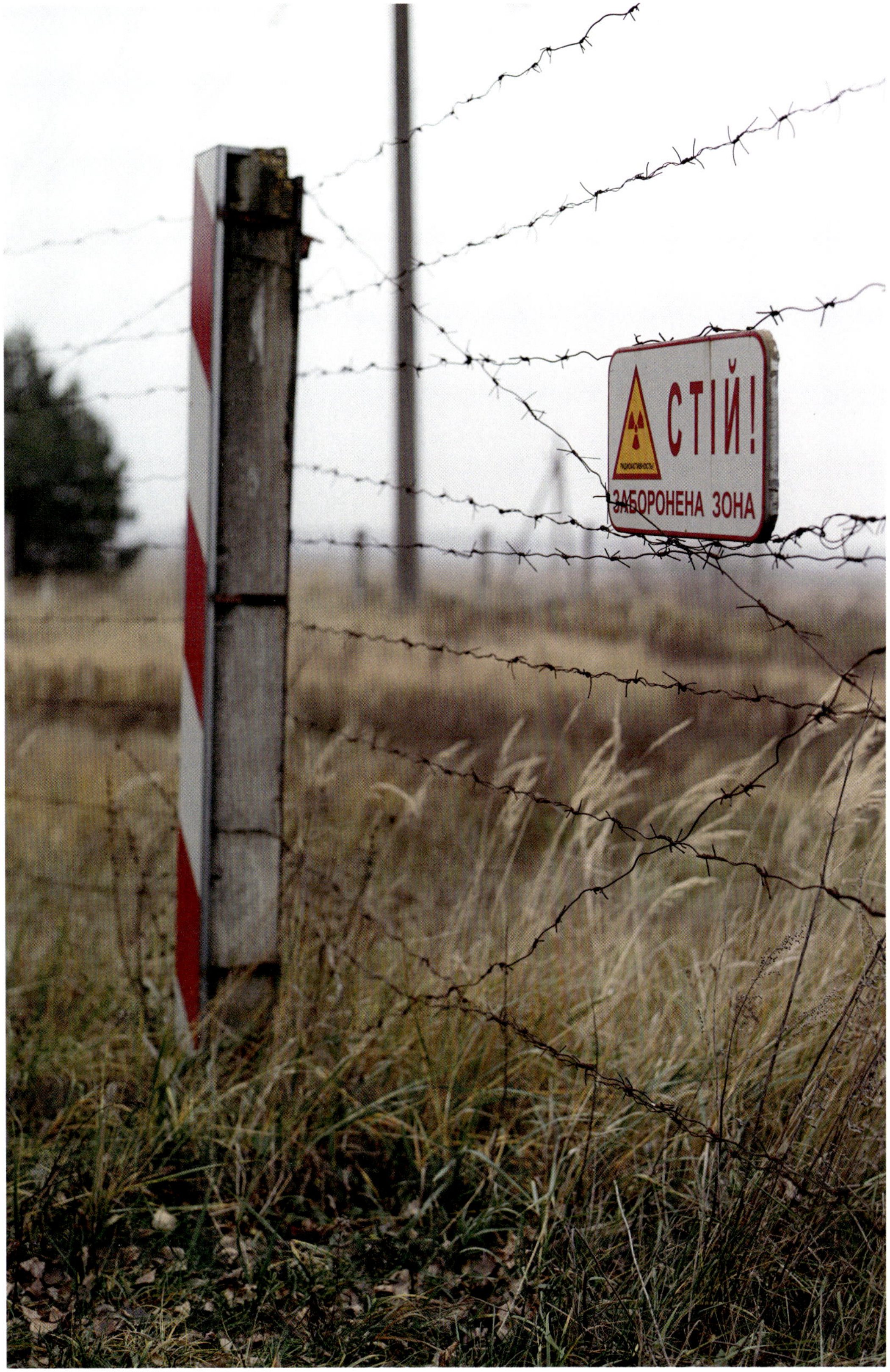
СТІЙ!
ЗАБОРОНЕНА ЗОНА

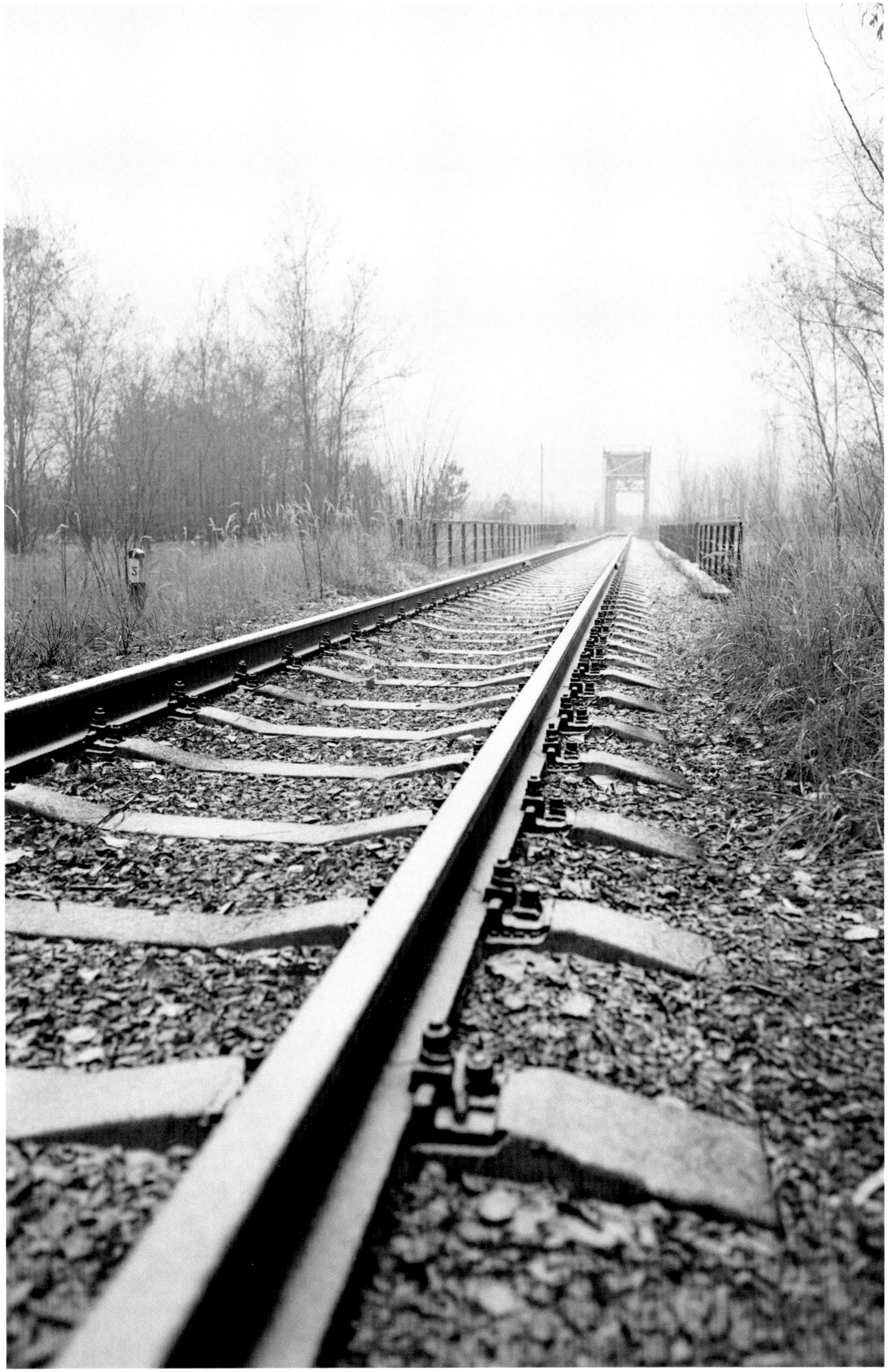

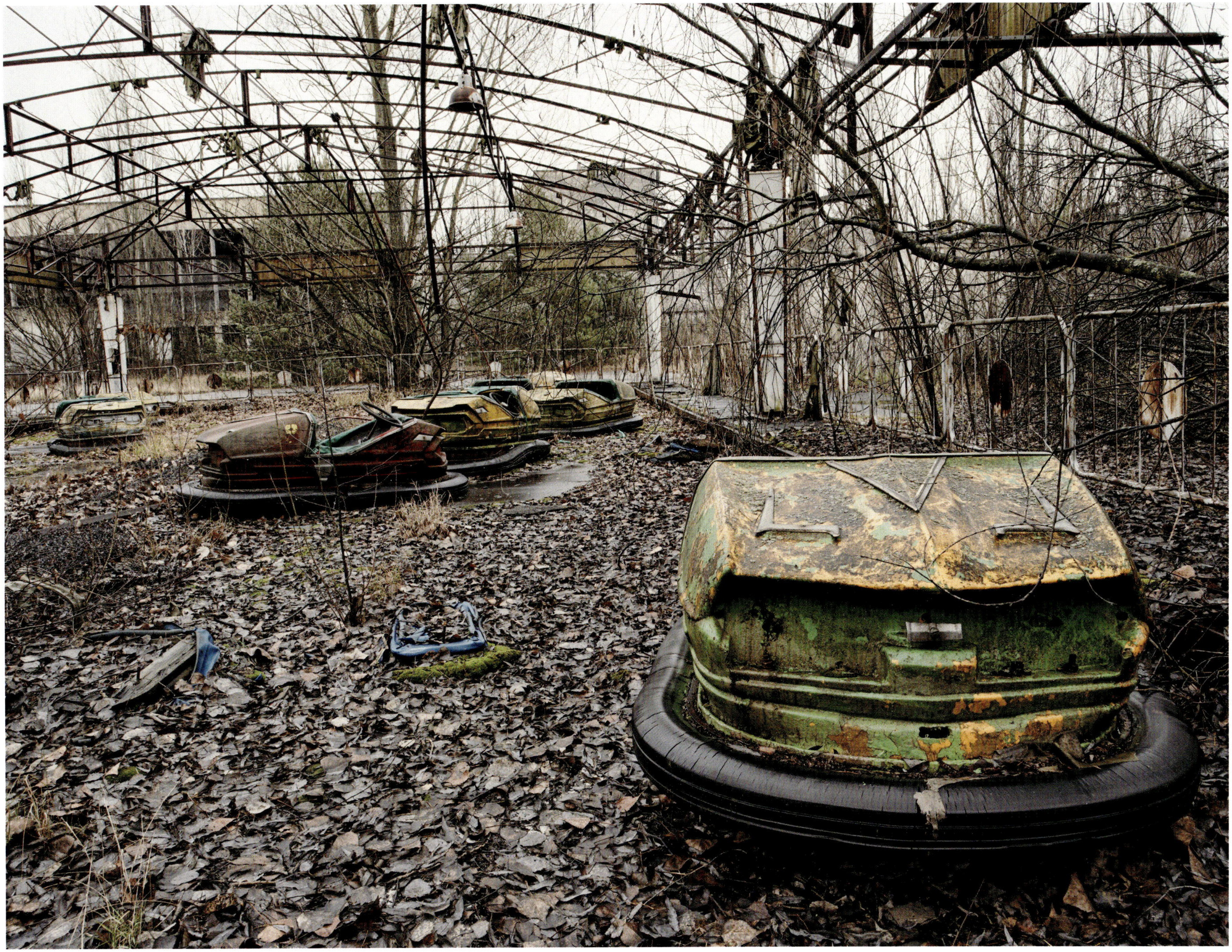

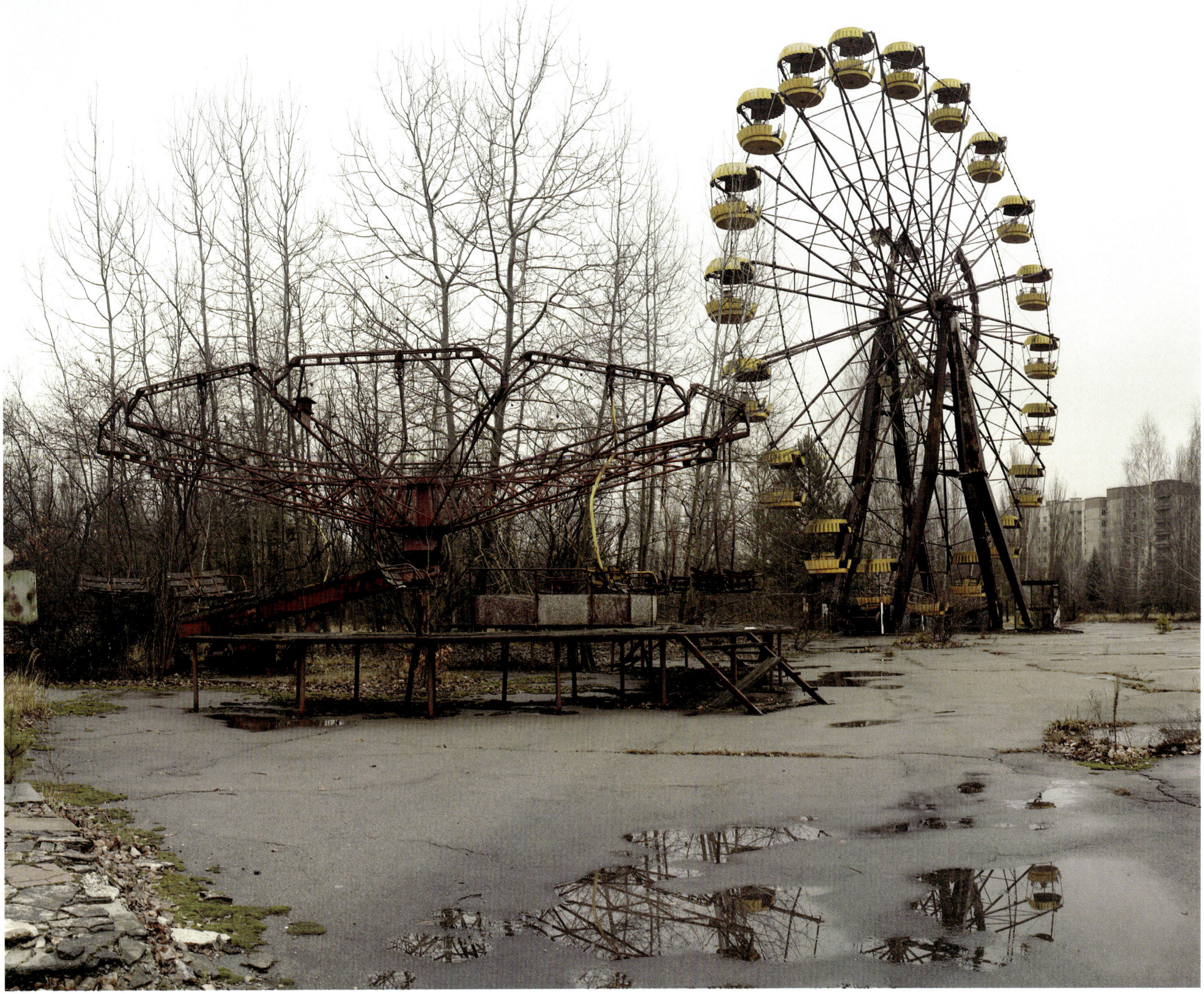

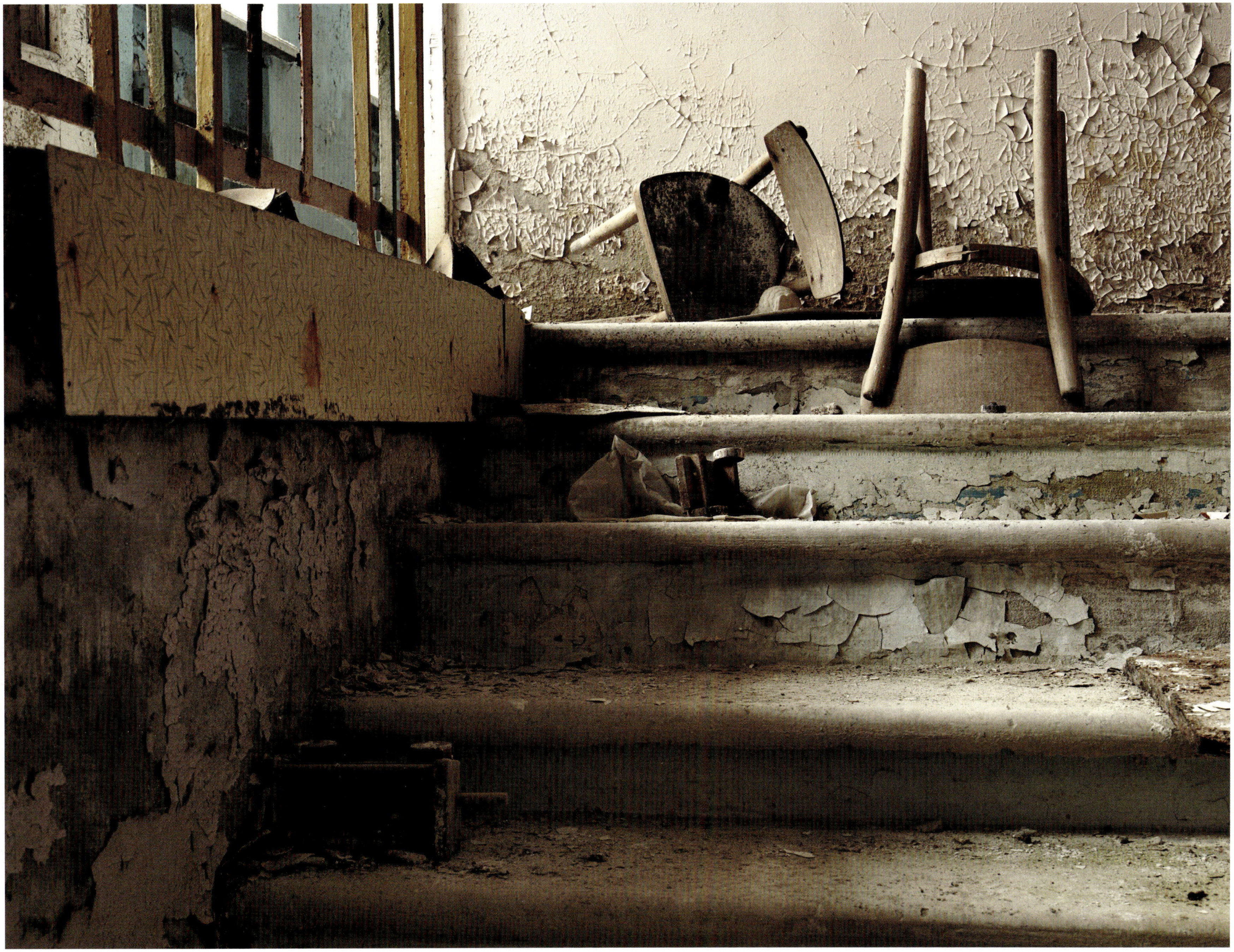

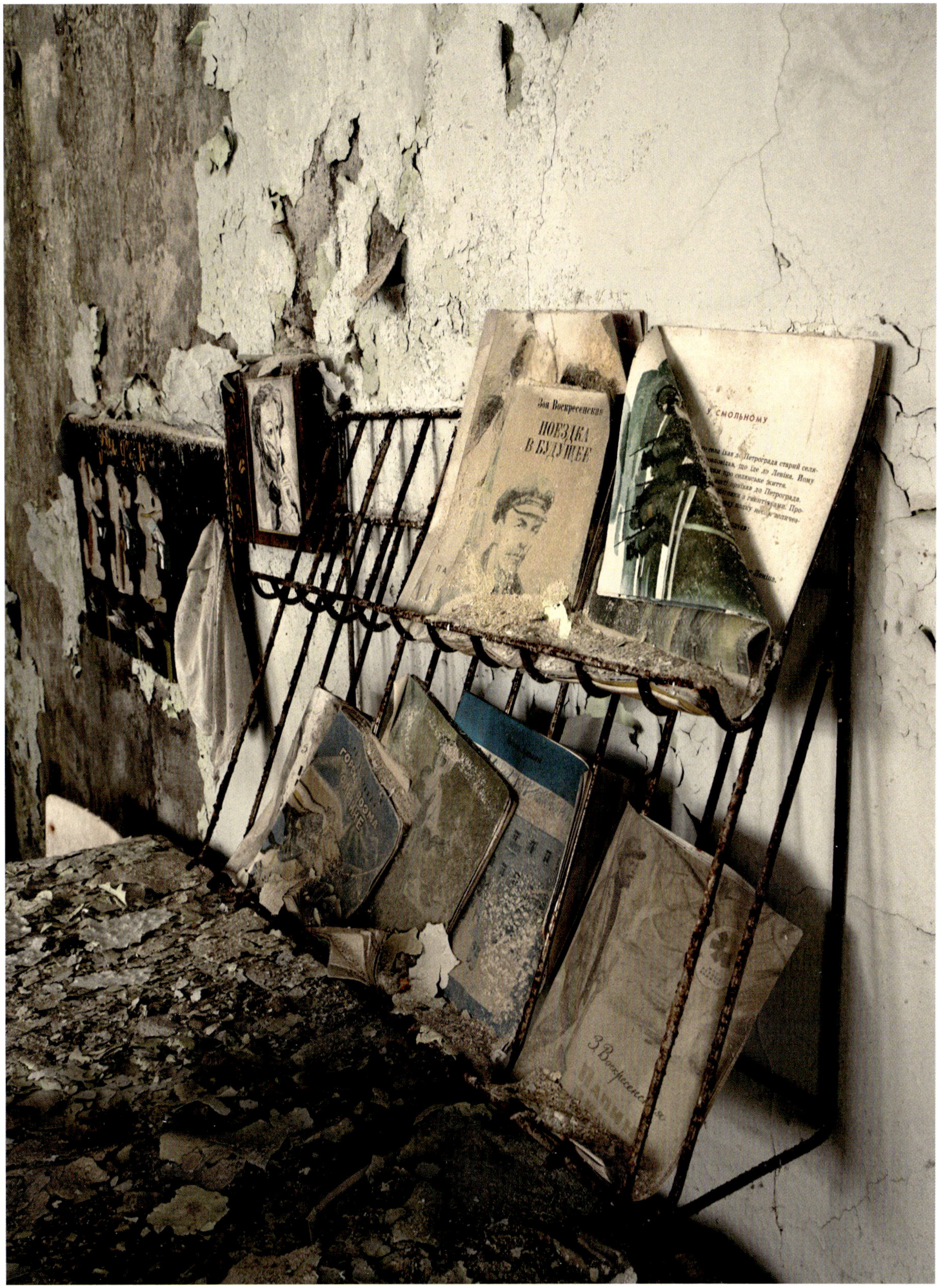
Зоя Воскресенская
ПОЕЗДКА
В БУДУЩЕЕ
У СМОЛЬНОМУ

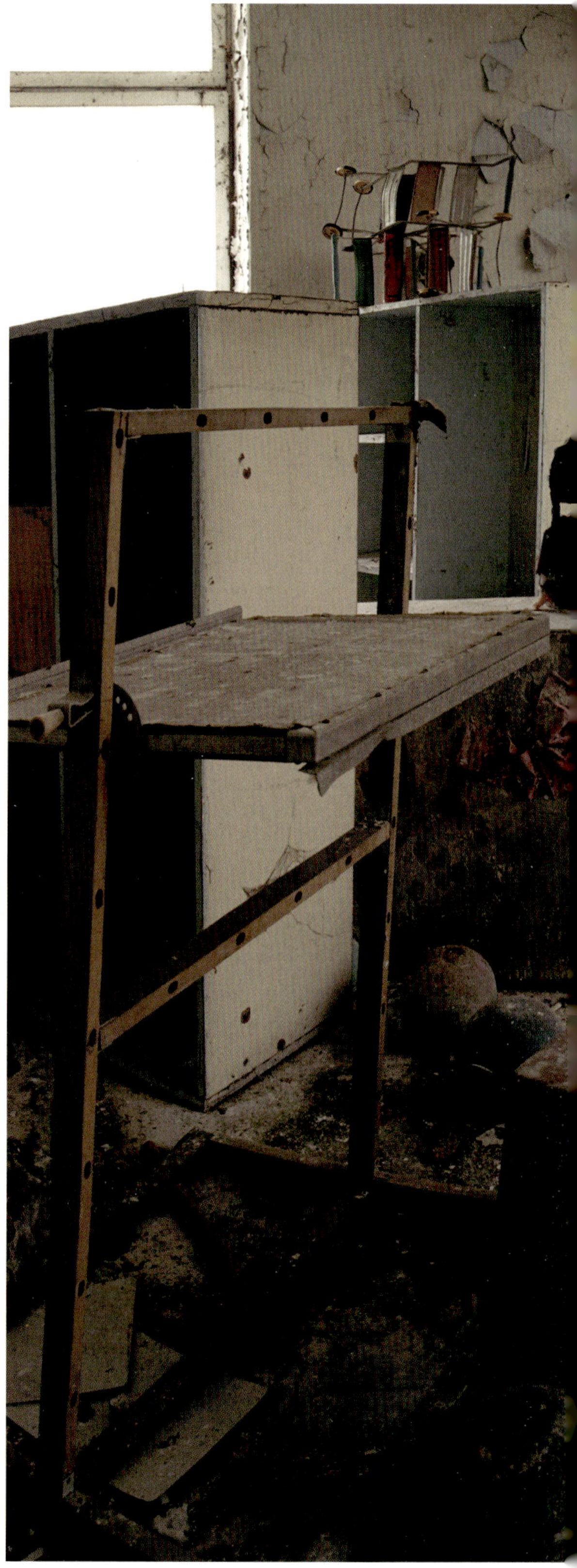

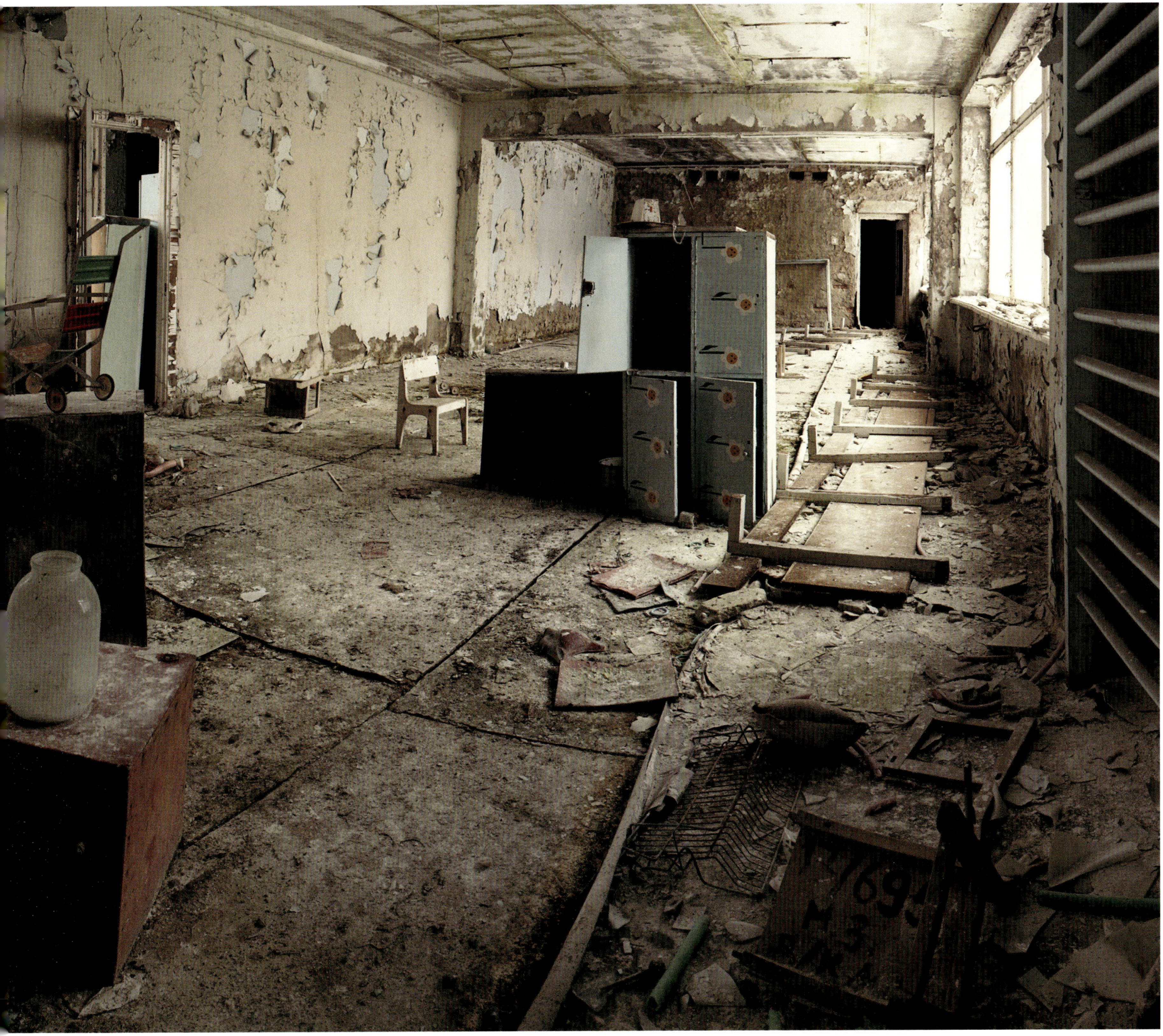

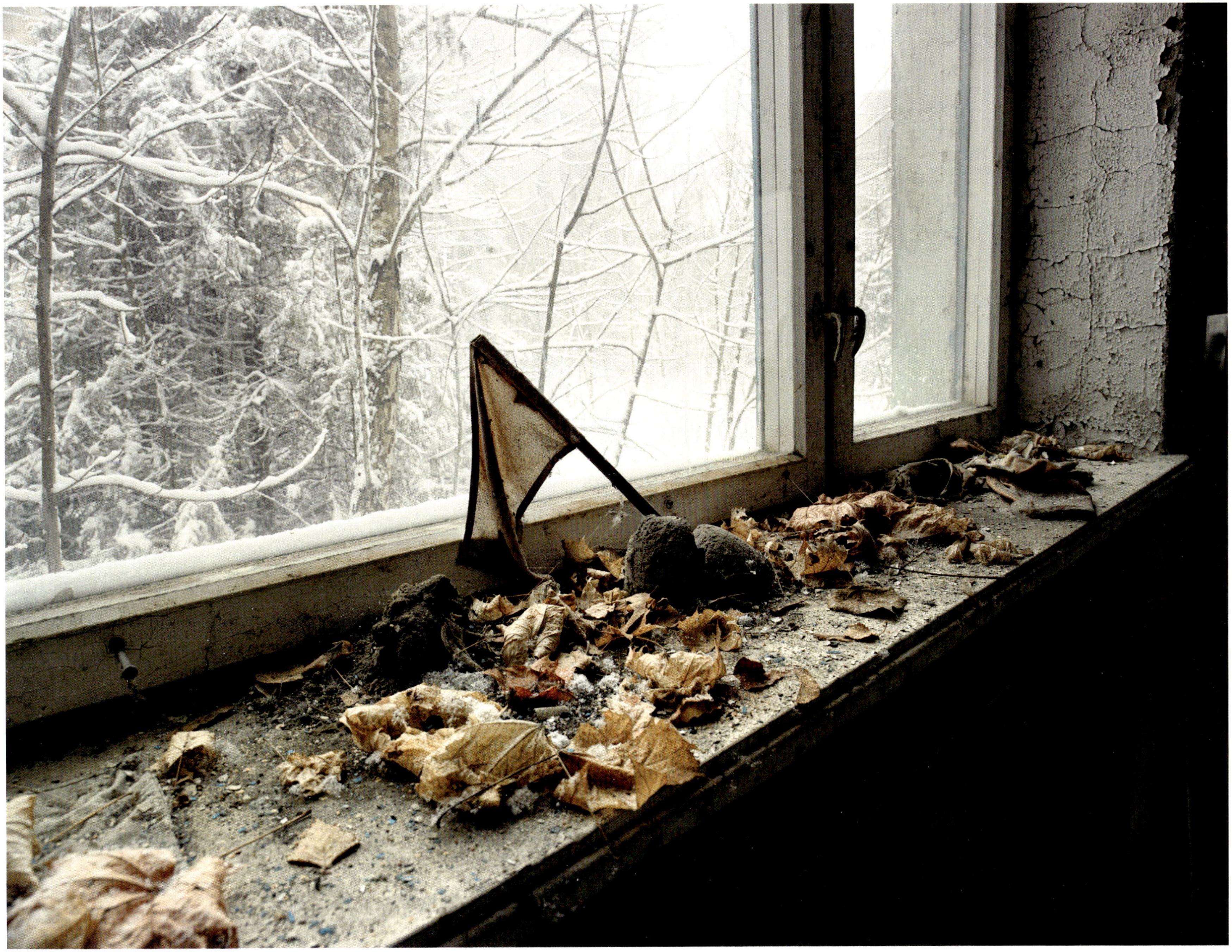

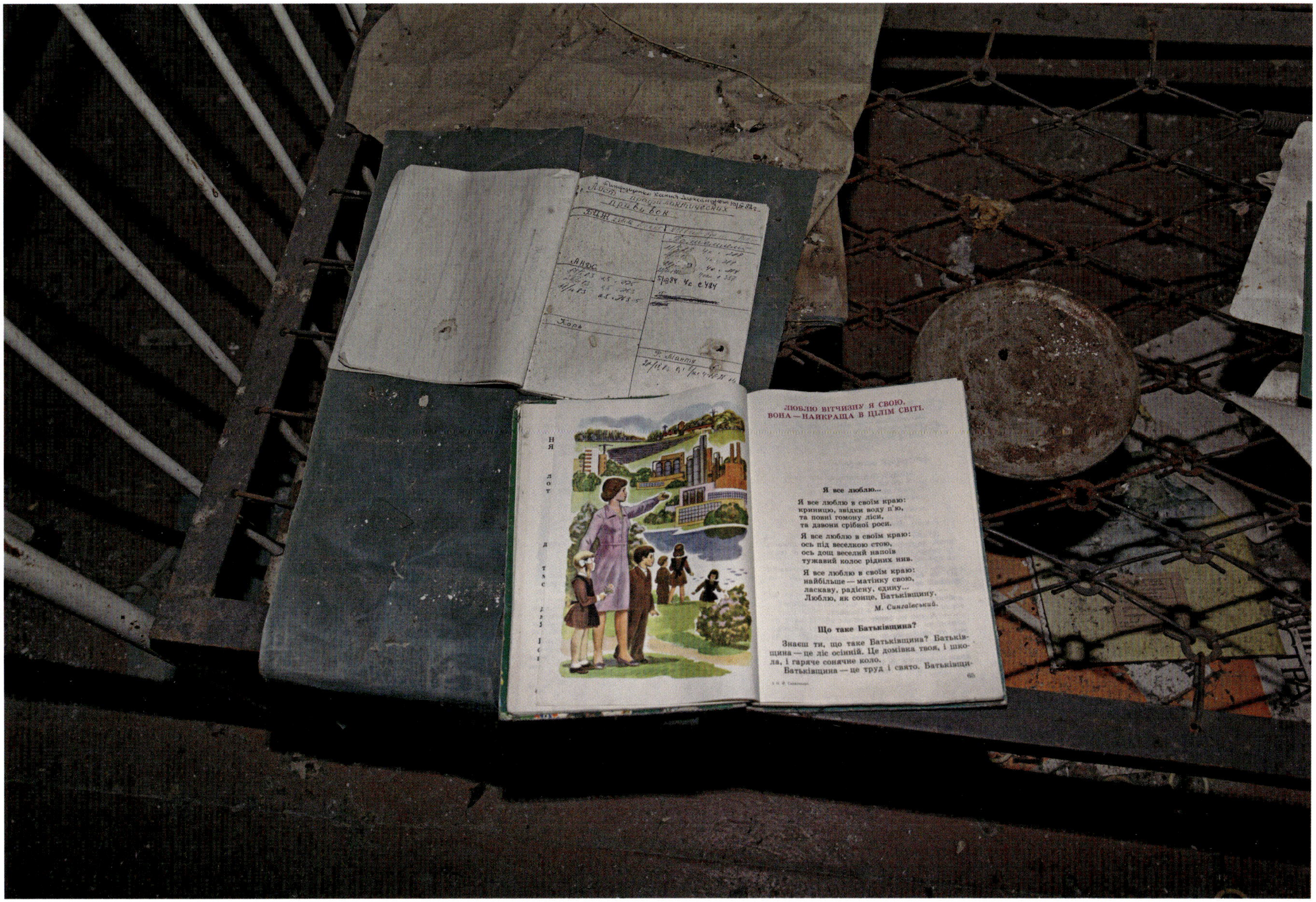
ЛЮБЛЮ ВІТЧИЗНУ Я СВОЮ,
ВОНА — НАЙКРАЩА В ЦІЛІМ СВІТІ.
Я все люблю...
Я все люблю в своїм краю:
криницю, звідки воду п'ю,
та повні гомону ліси,
та дзвони срібної роси.
Я все люблю в своїм краю:
ось під веселкою стою,
ось дощ веселий напоїв
тужавий колос рідних нив.
Я все люблю в своїм краю:
найбільше — матінку свою,
ласкаву, радісну, єдину...
Люблю, як сонце, Батьківщину.
М. Сингаївський.
Що таке Батьківщина?
Знаєш ти, що таке Батьківщина? Батьків-
щина — це ліс осінній. Це домівка твоя, і шко-
ла, і гаряче сонячне коло.
Батьківщина — це труд і свято. Батьківщи-

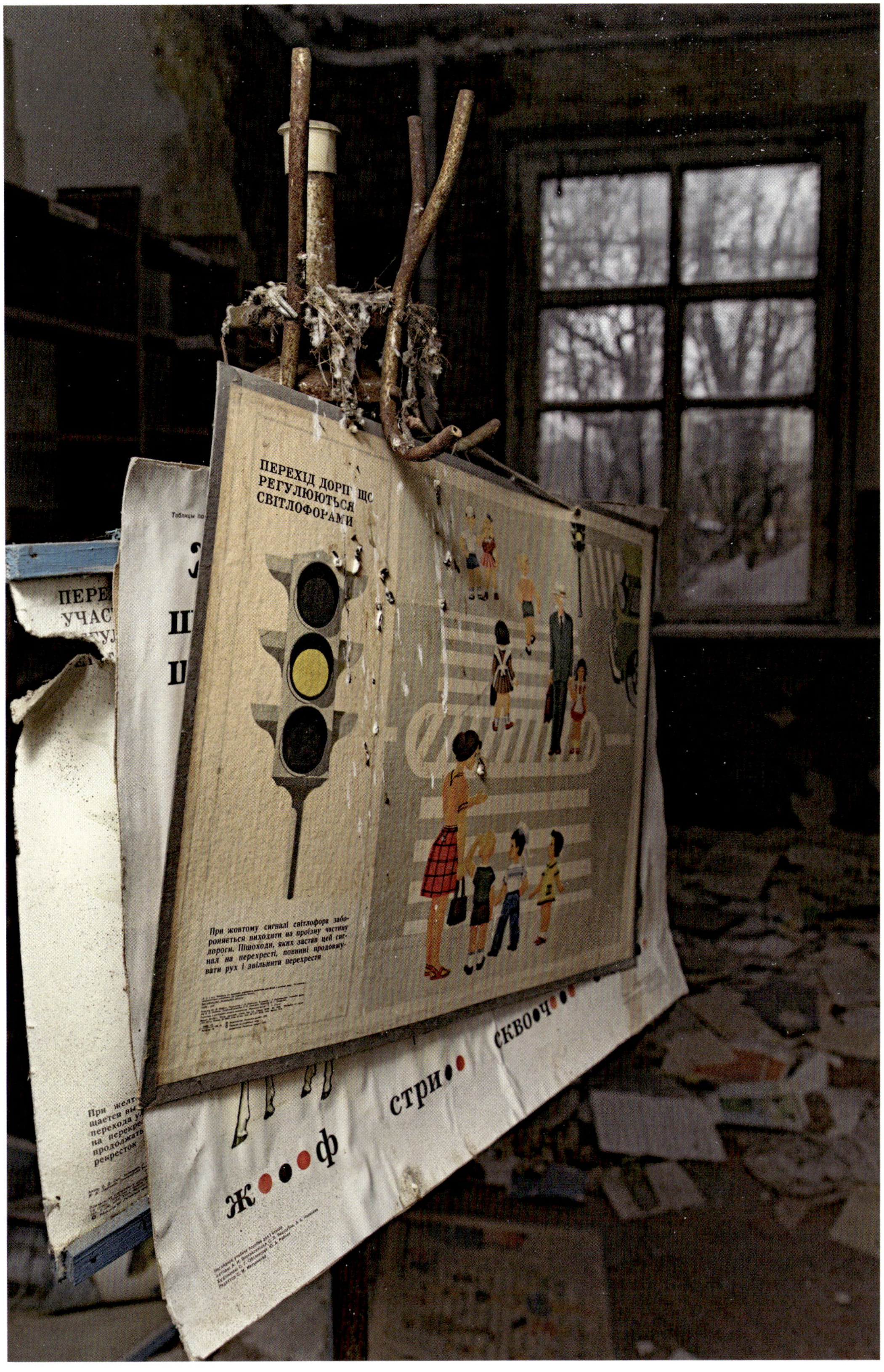

ПЕРЕХІД ДОРІГ ЩО
РЕГУЛЮЮТЬСЯ
СВІТЛОФОРАМИ
При жовтому сигналі світлофора забороняється виходити на проїзну частину дороги. Пішоходи, яких застав цей сигнал на перехресті, повинні продовжувати рух і звільнити перехрестя
Таблицы по
стри
сквоч
ж ф

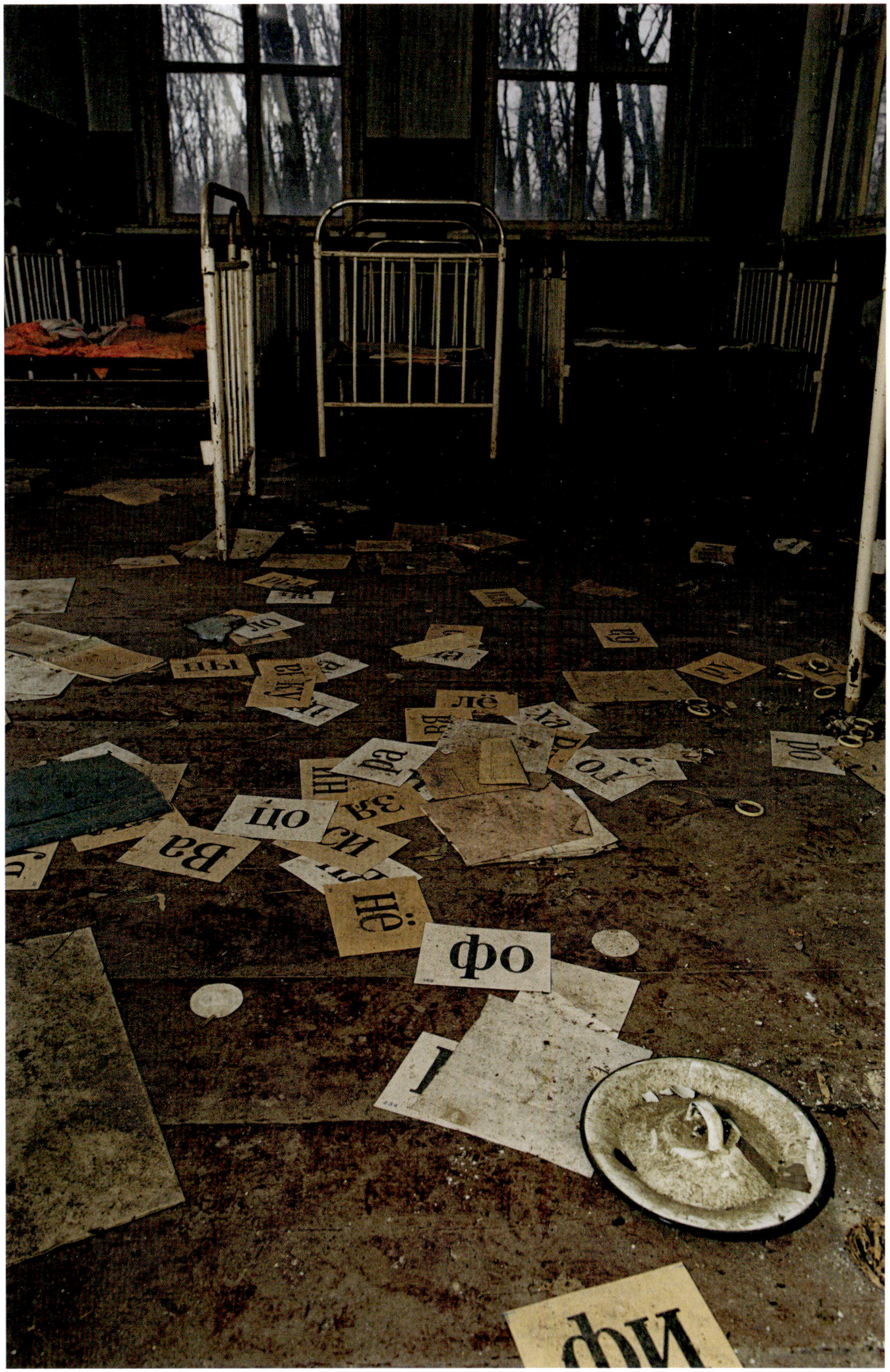
фо
нё
лё
фи

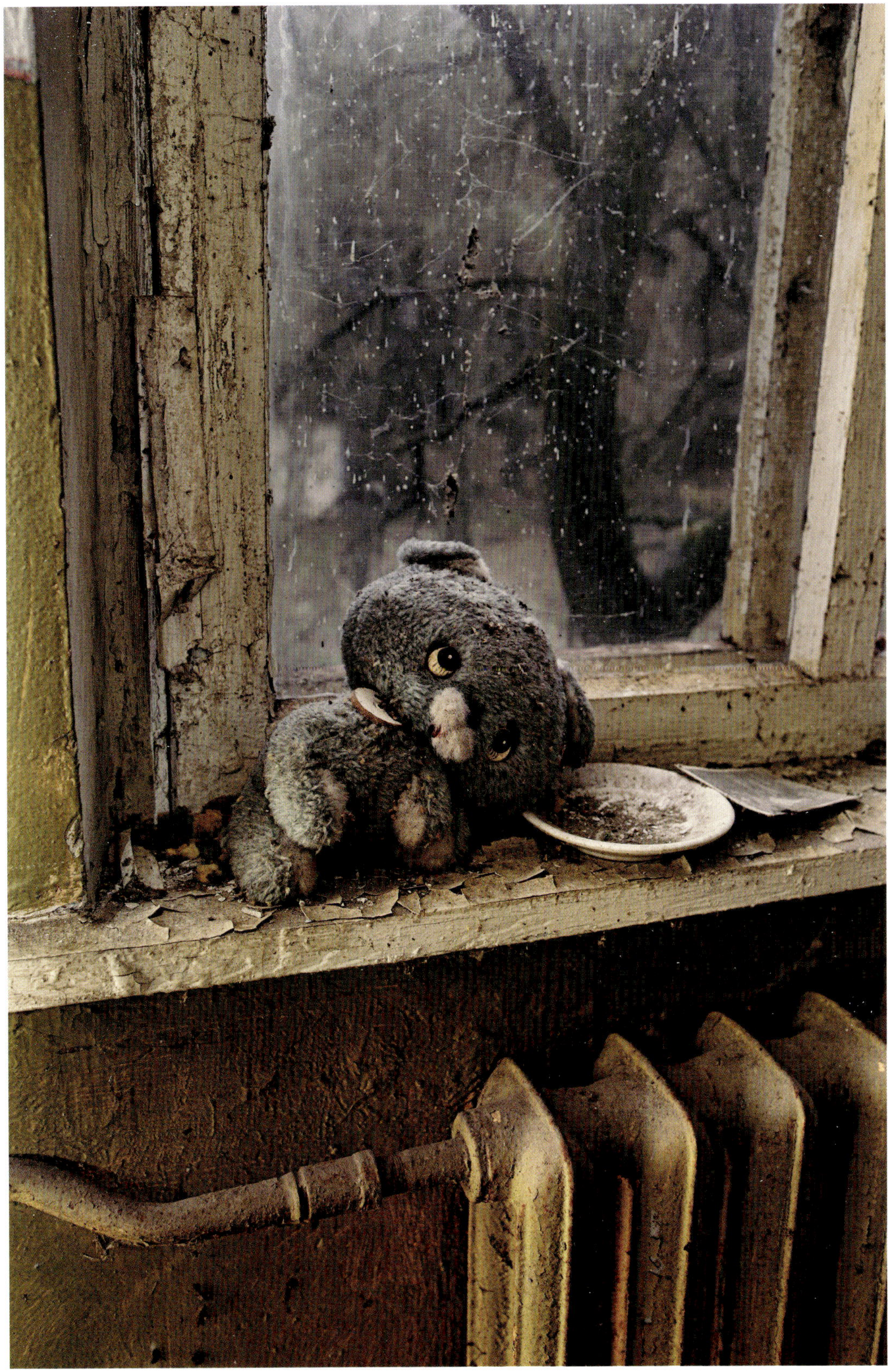

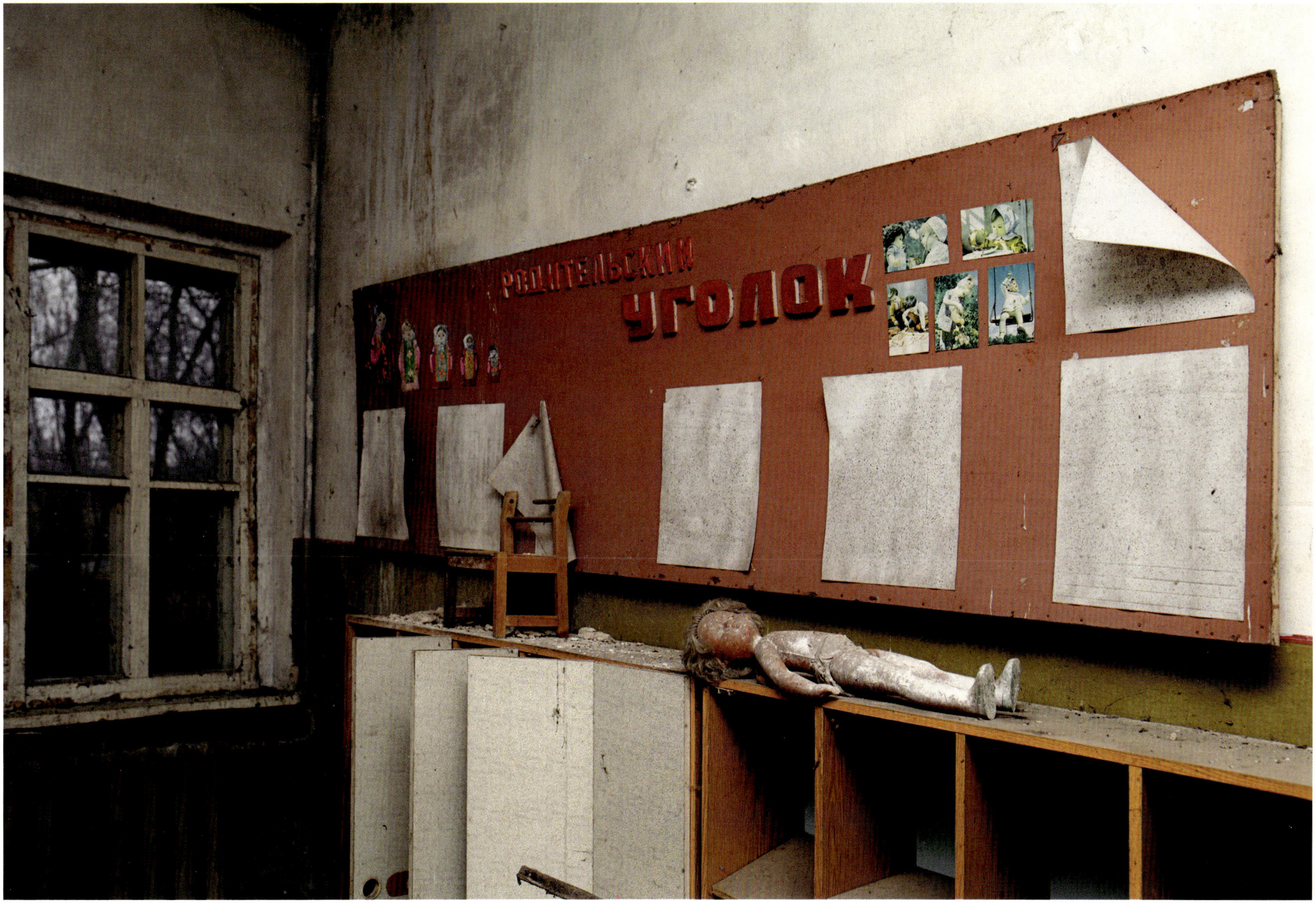
РОДИТЕЛЬСКИЙ
УГОЛОК

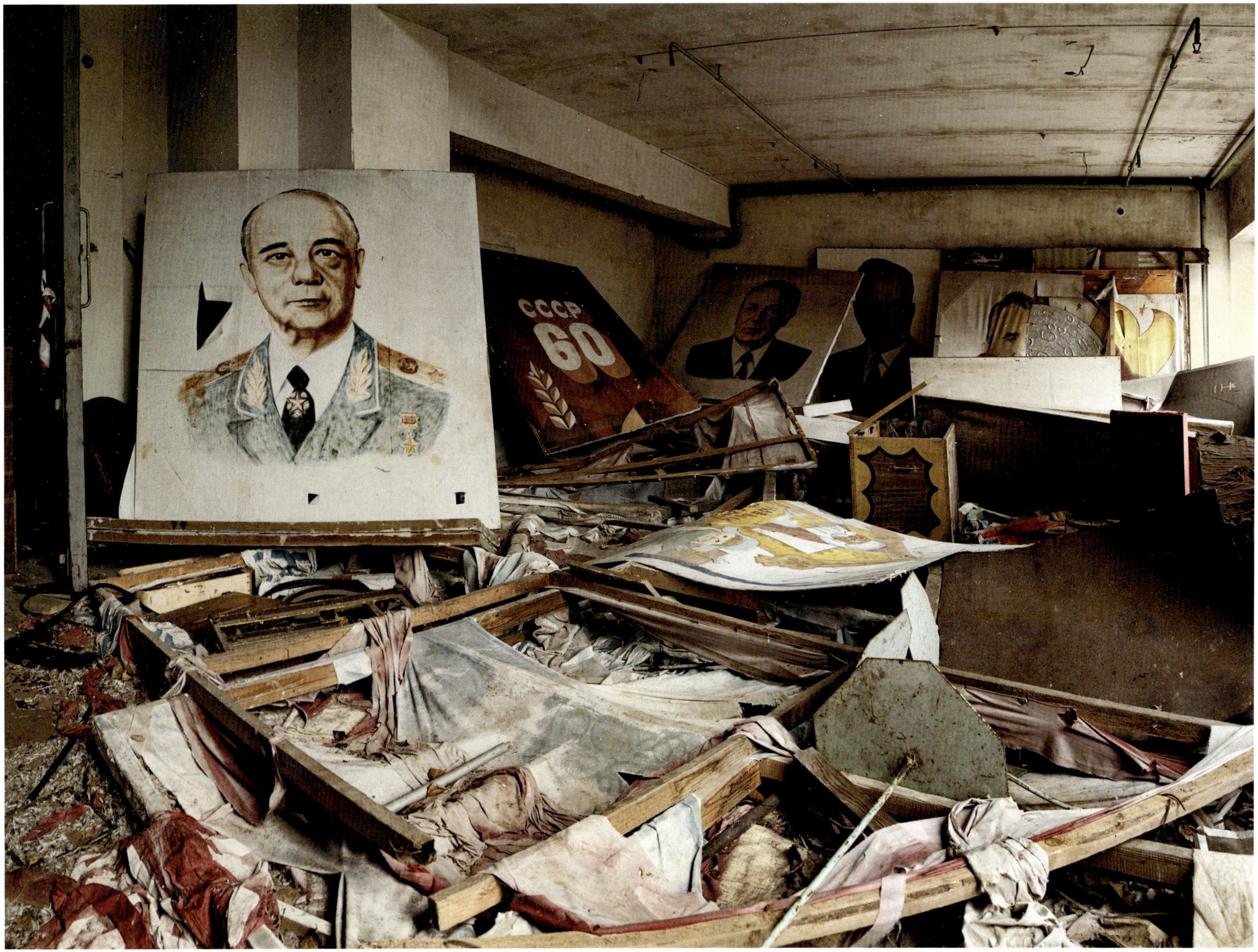
CCCP
60

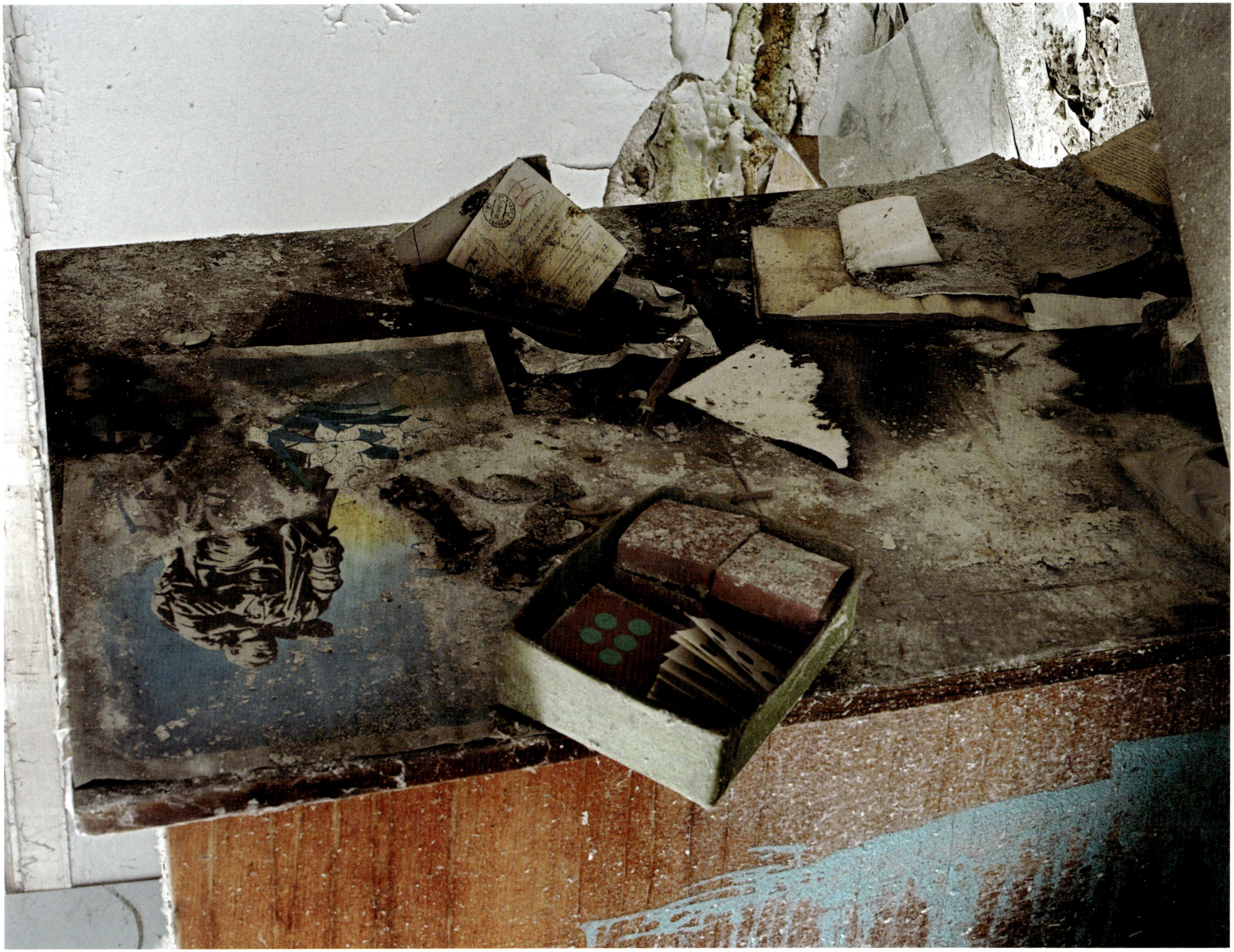

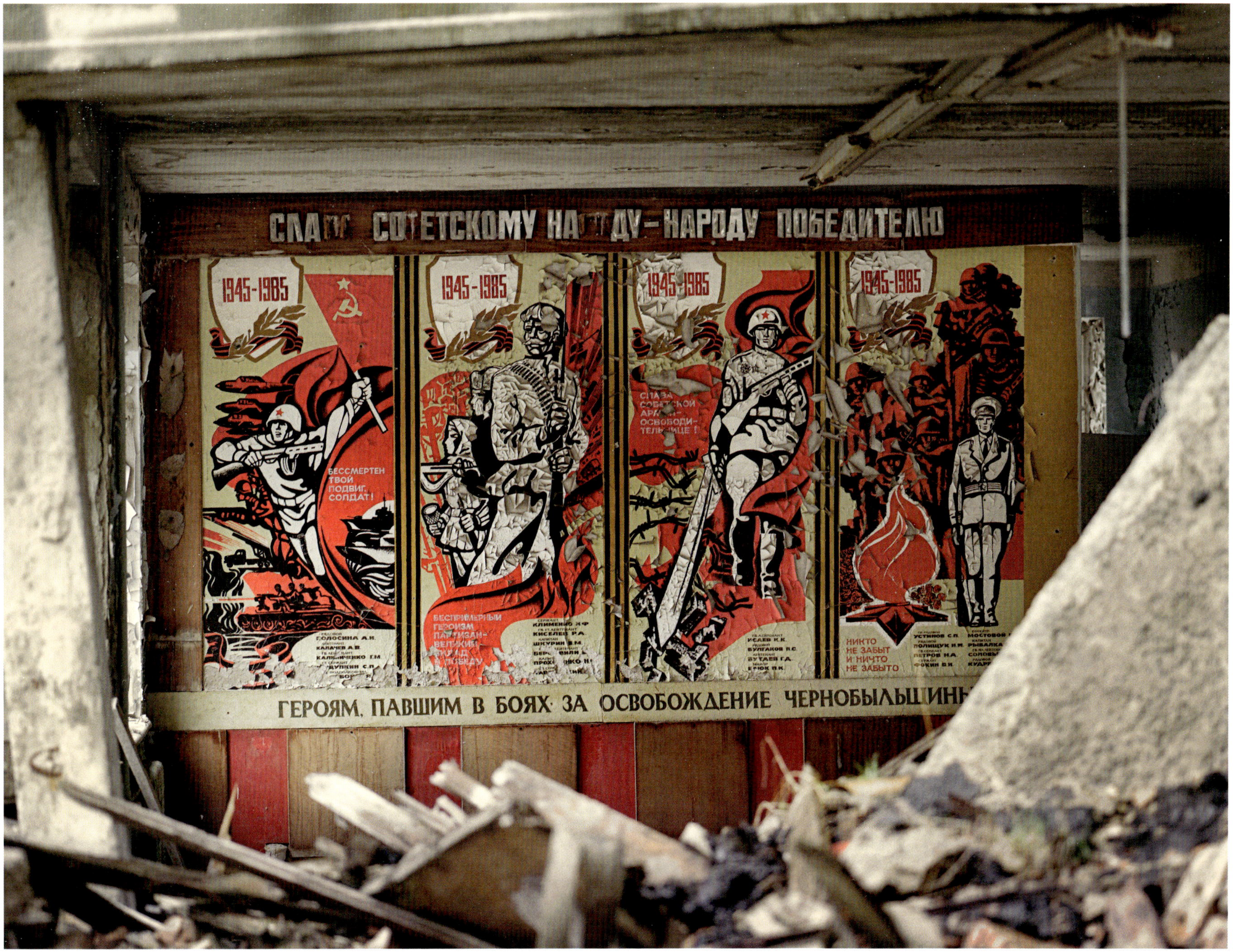

СЛА СО ЕТСКОМУ НА ДУ–НАРОДУ ПОБЕДИТЕЛЮ
1945–1985
1945–1985
1945–1985
1945–1985
БЕССМЕРТЕН ТВОЙ ПОДВИГ, СОЛДАТ!
НИКТО НЕ ЗАБЫТ И НИЧТО НЕ ЗАБЫТО
ГЕРОЯМ, ПАВШИМ В БОЯХ ЗА ОСВОБОЖДЕНИЕ ЧЕРНОБЫЛЬЩИН

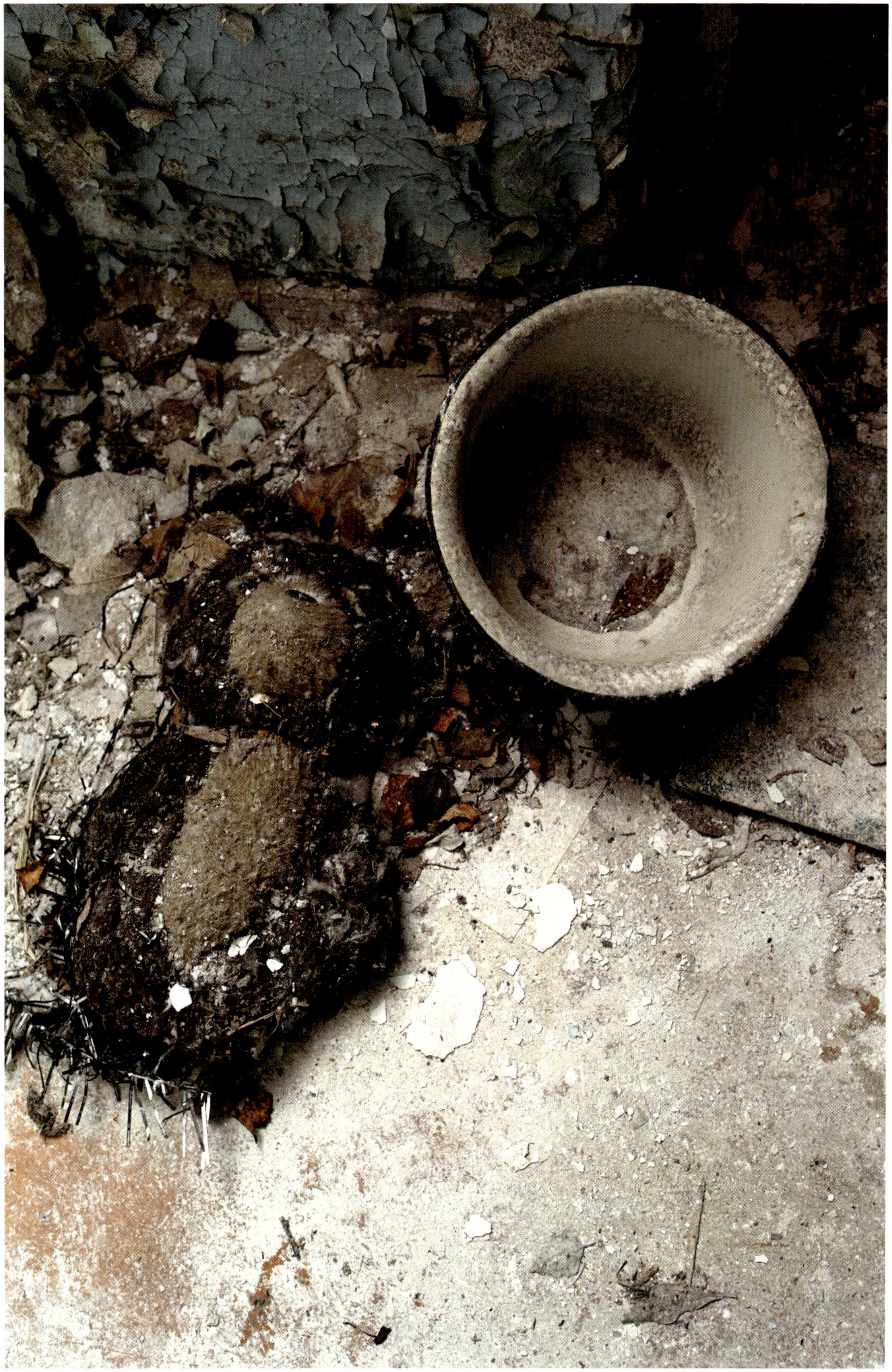

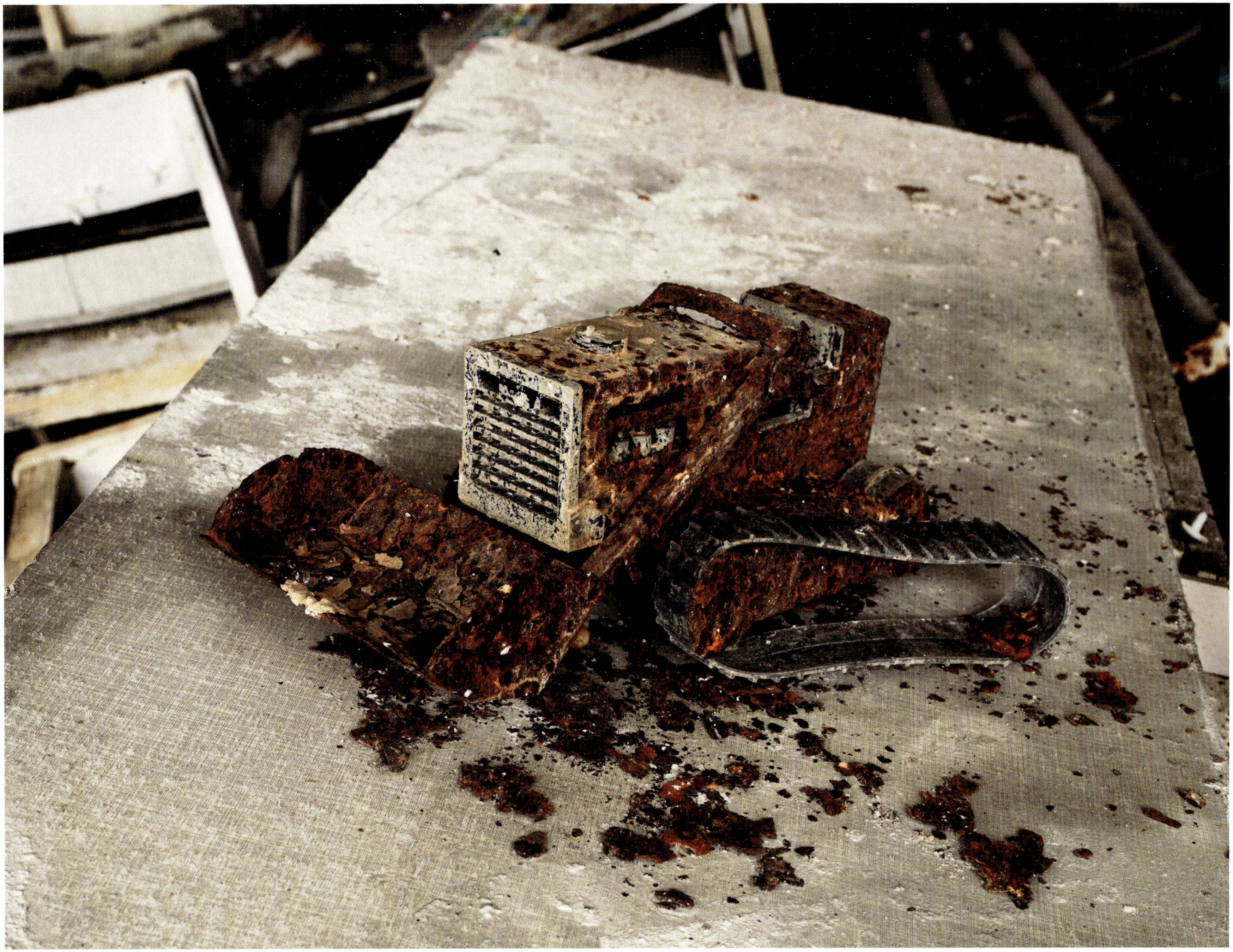

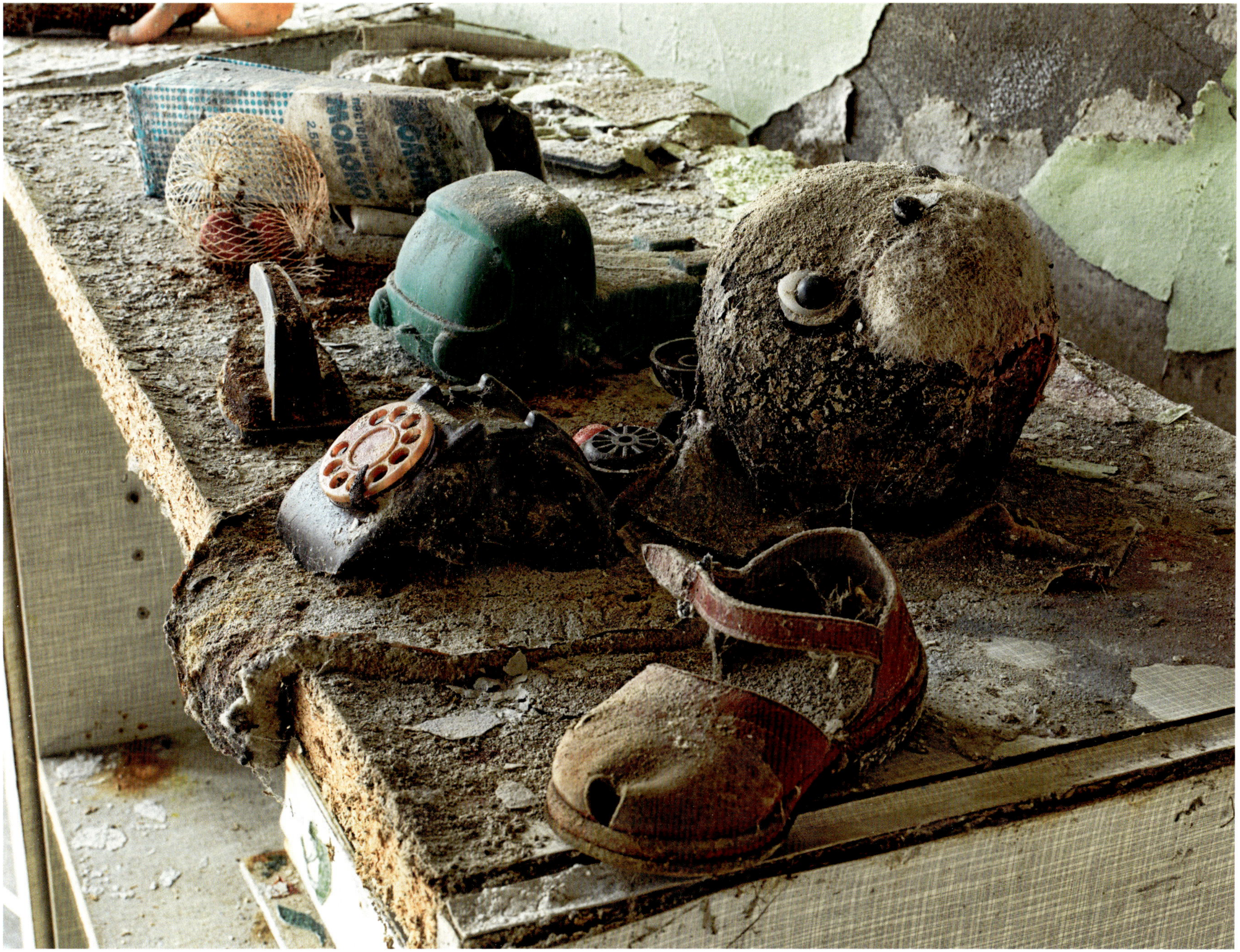
МОЛОКО

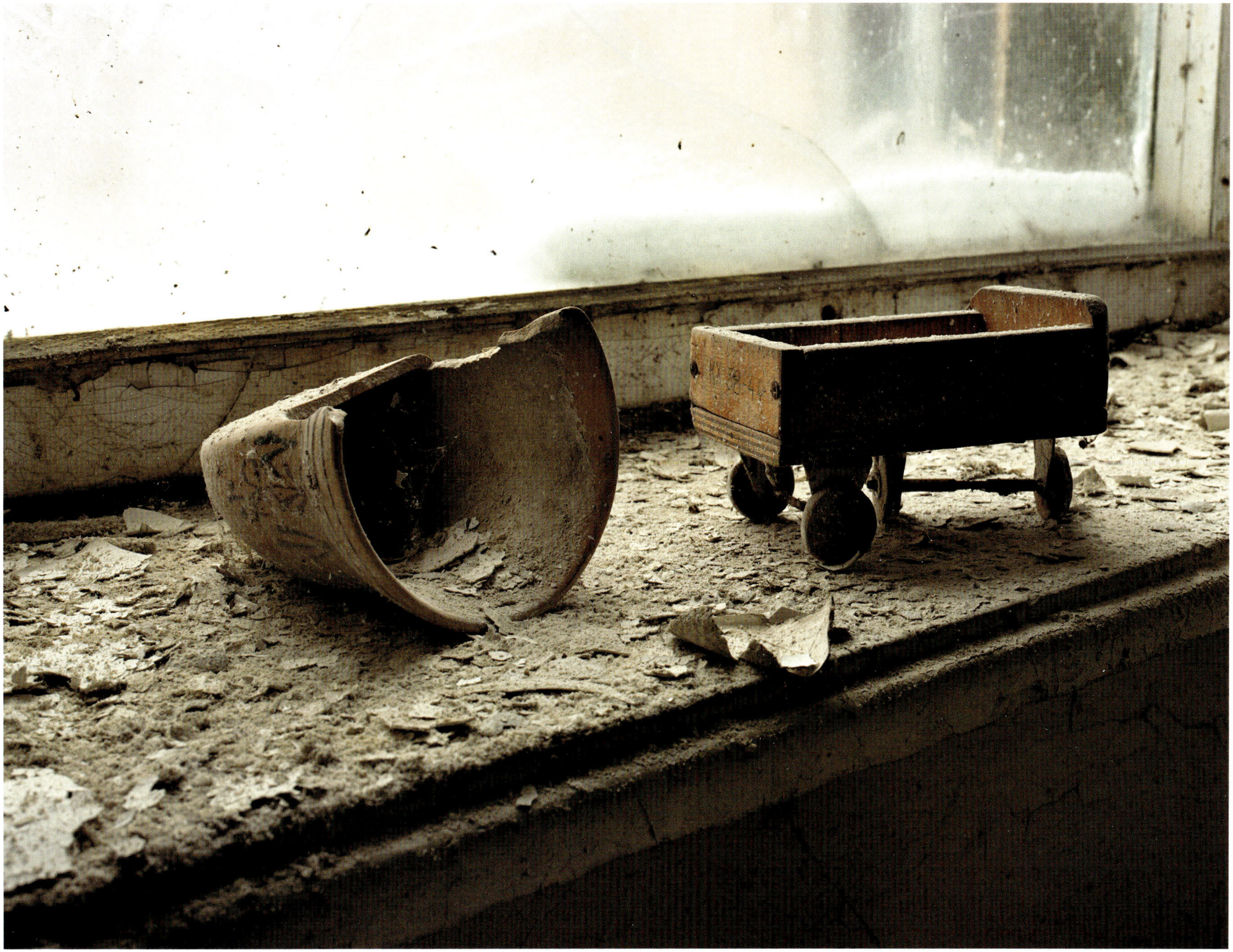

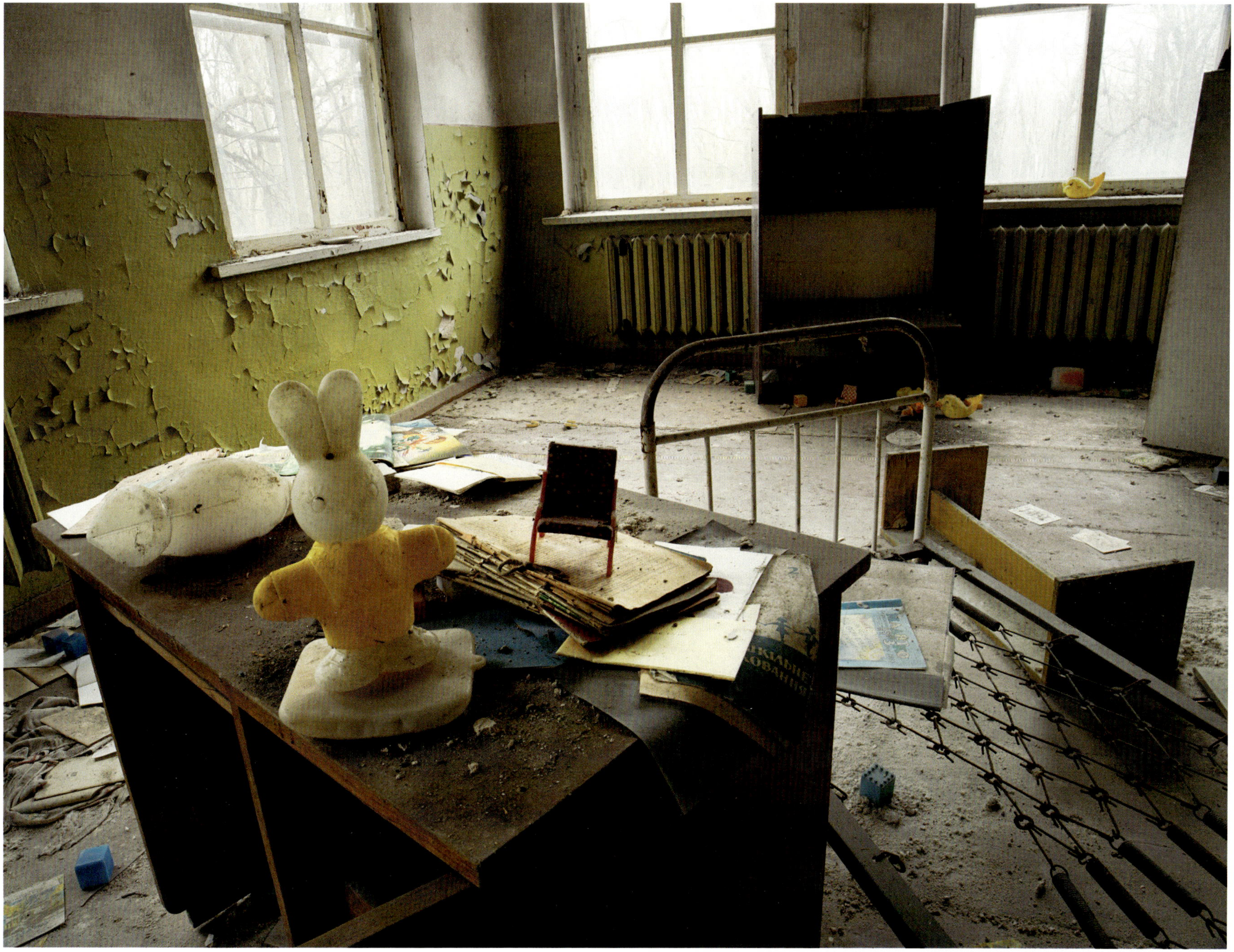

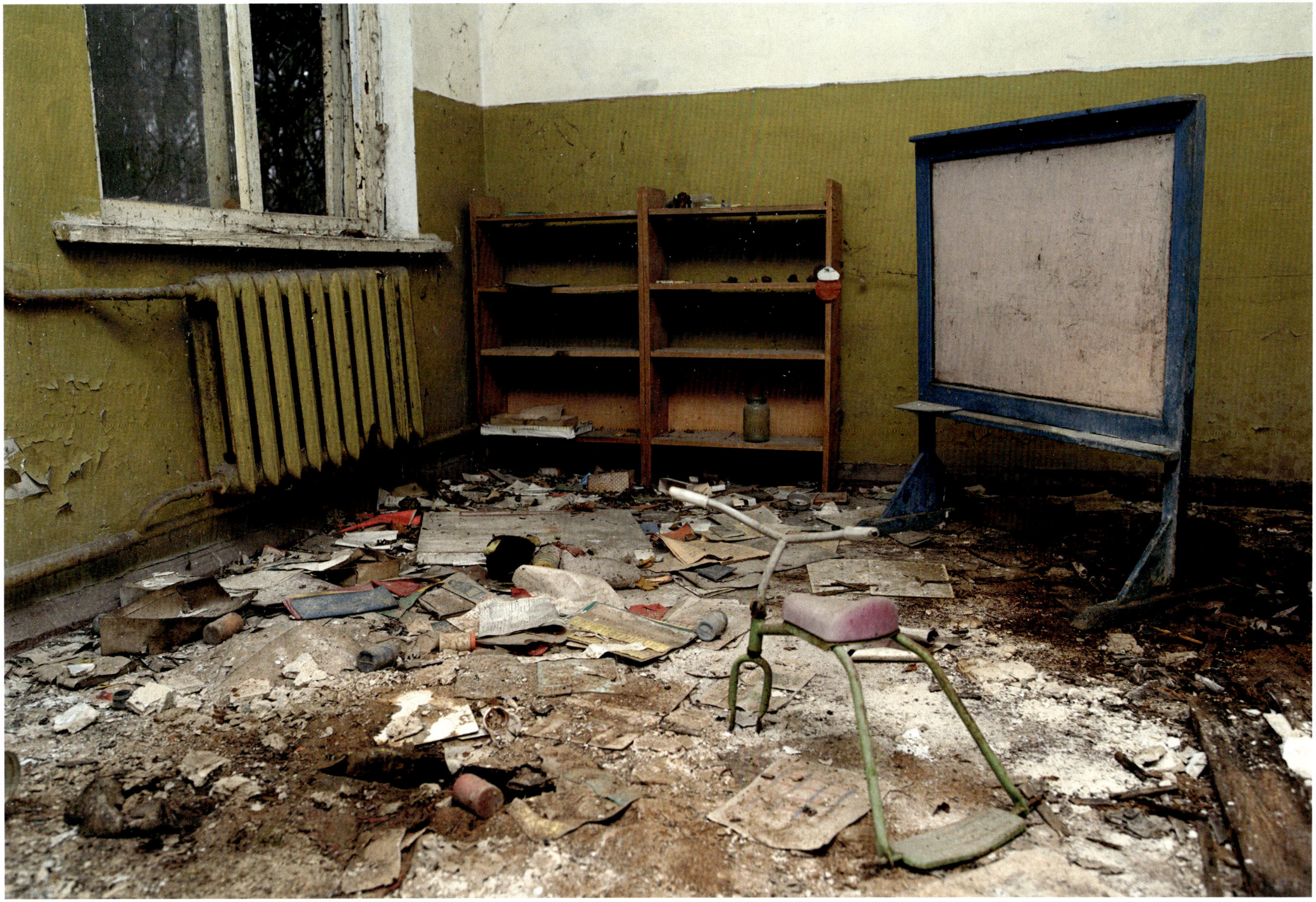

60

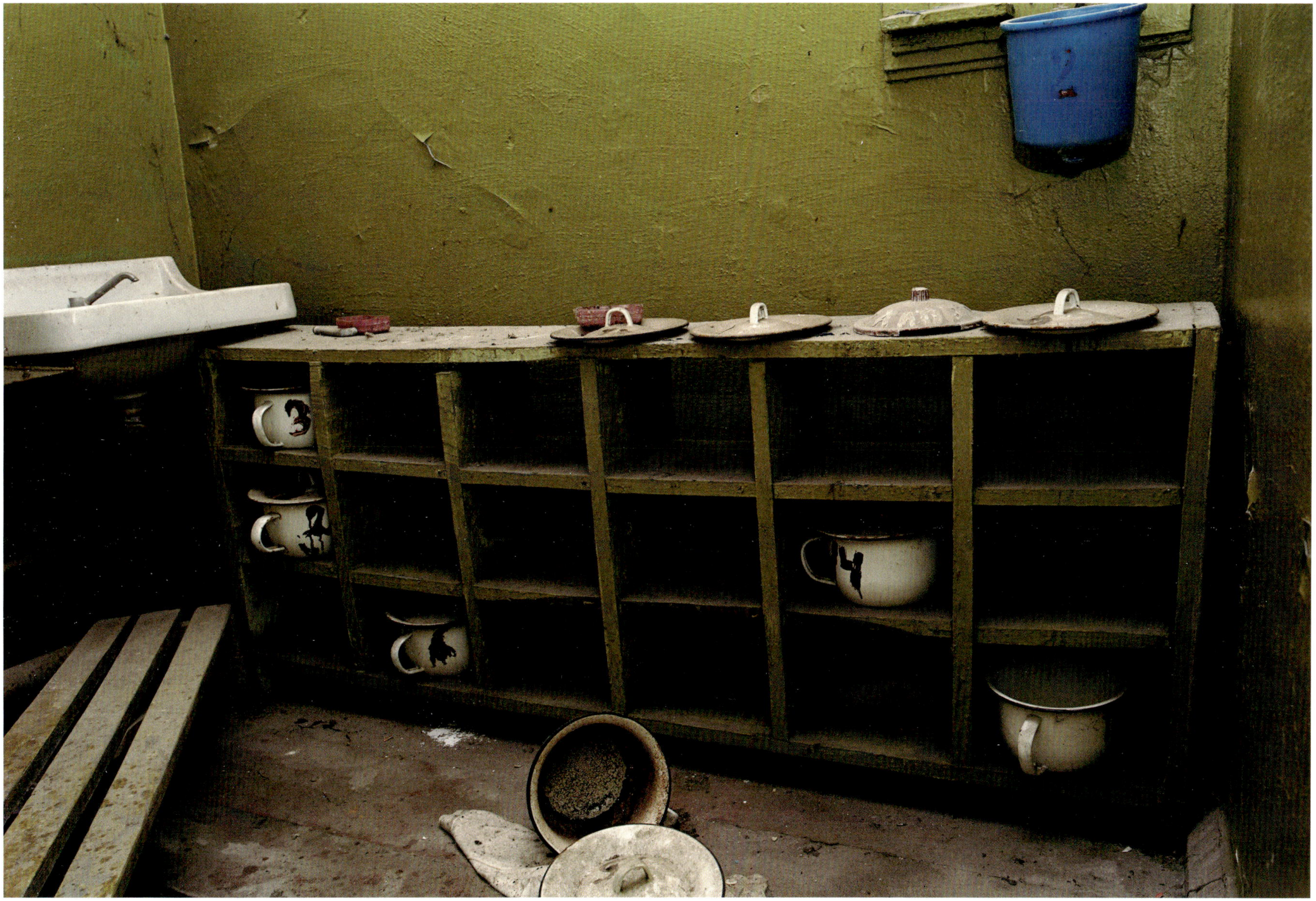

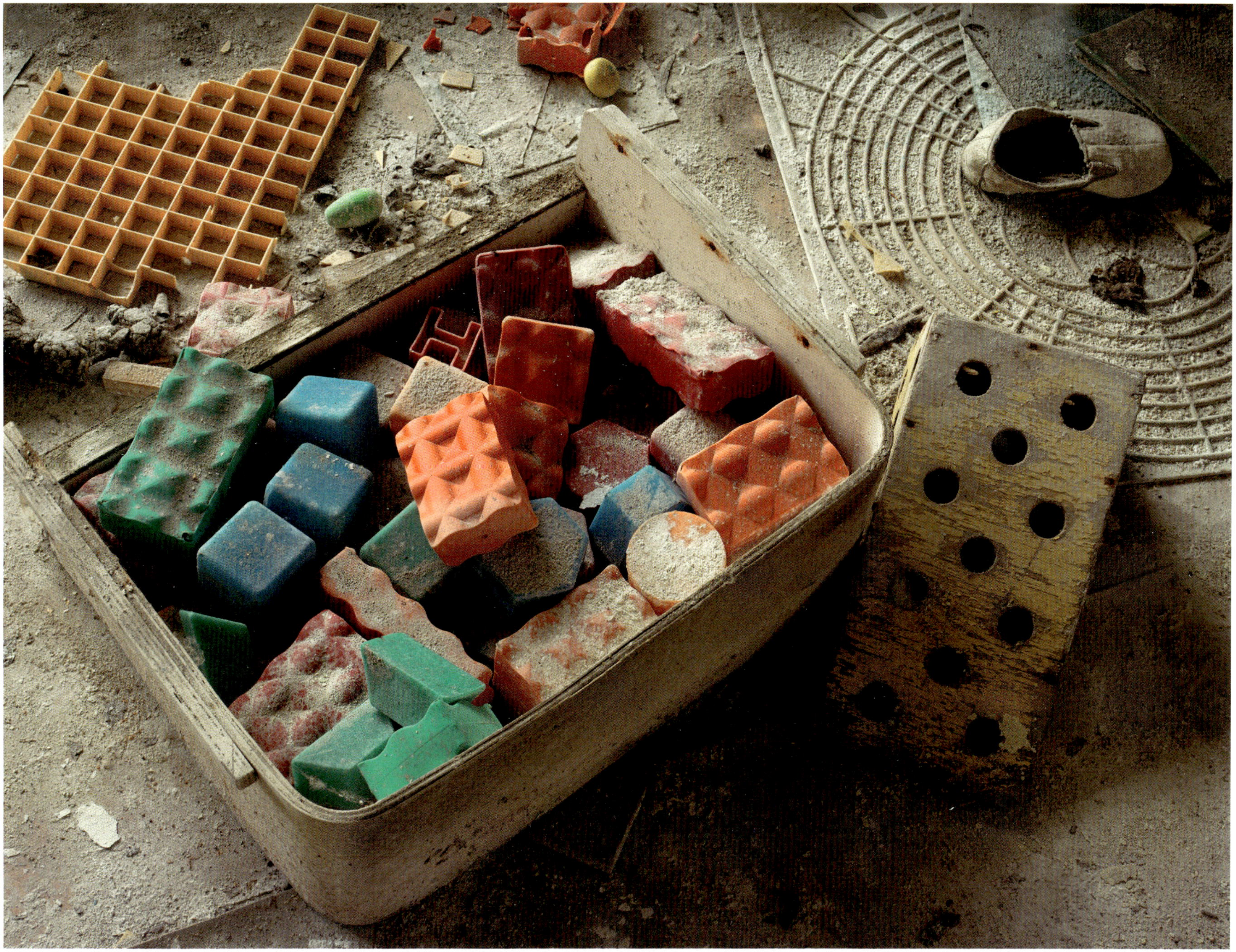

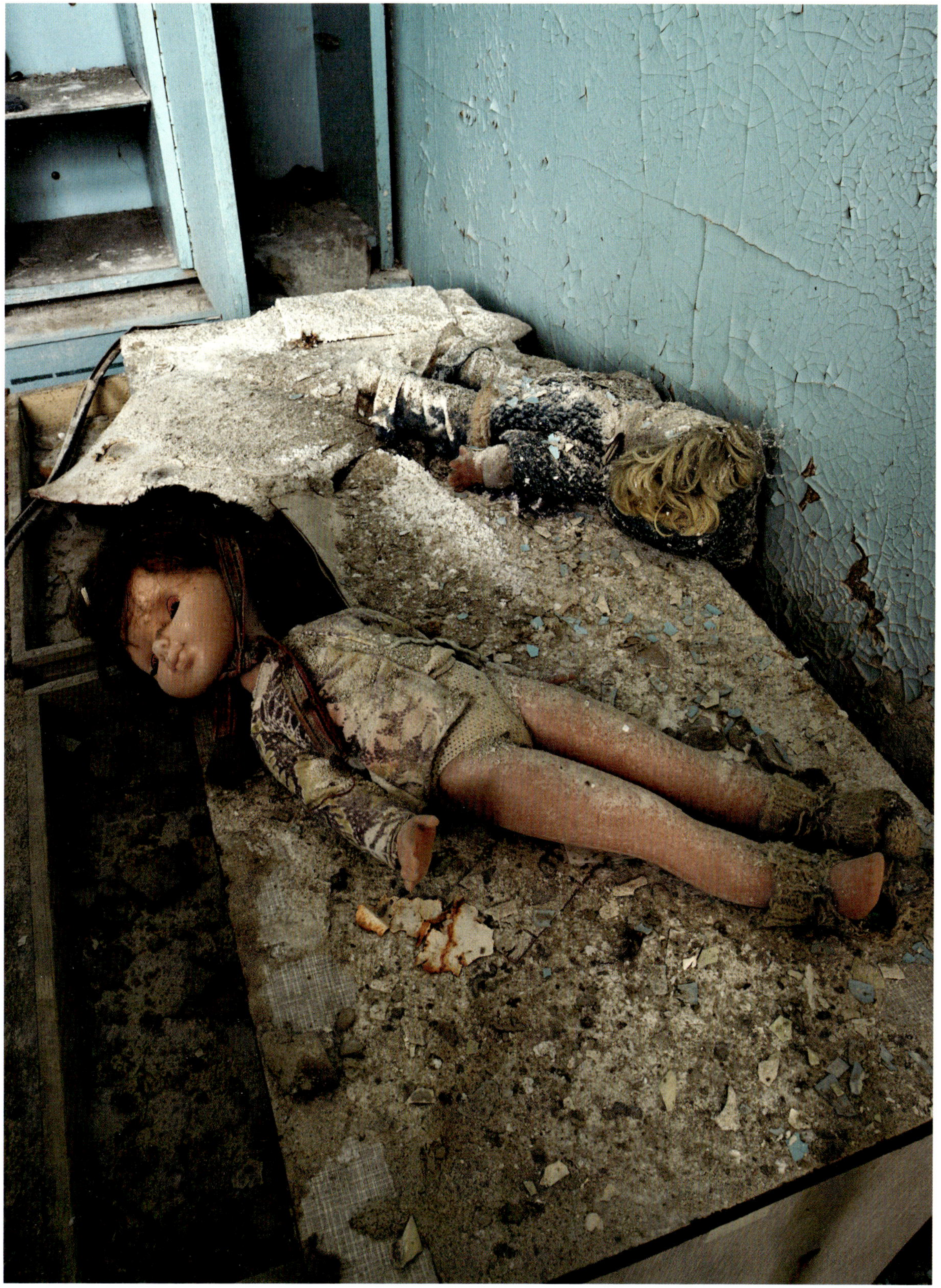

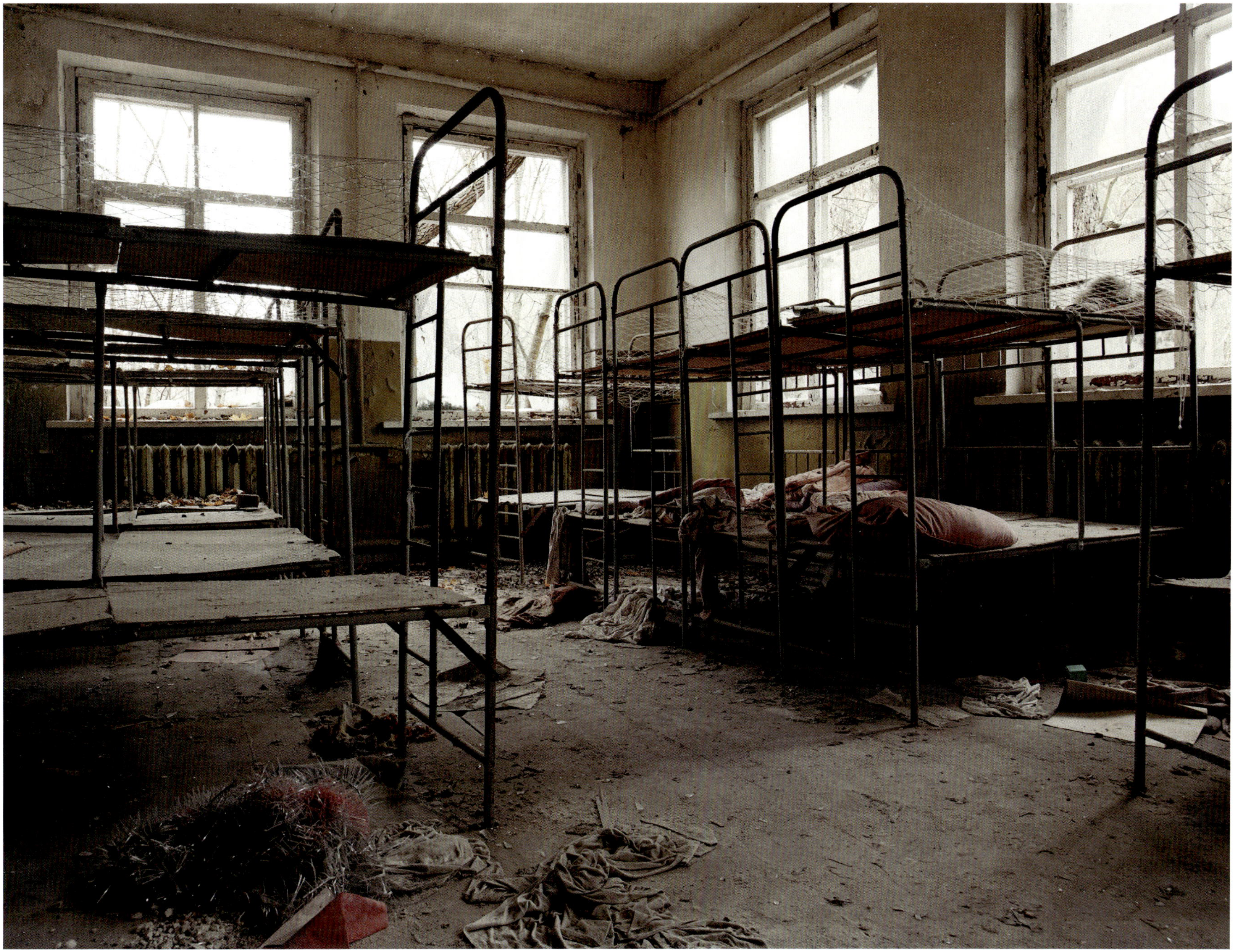

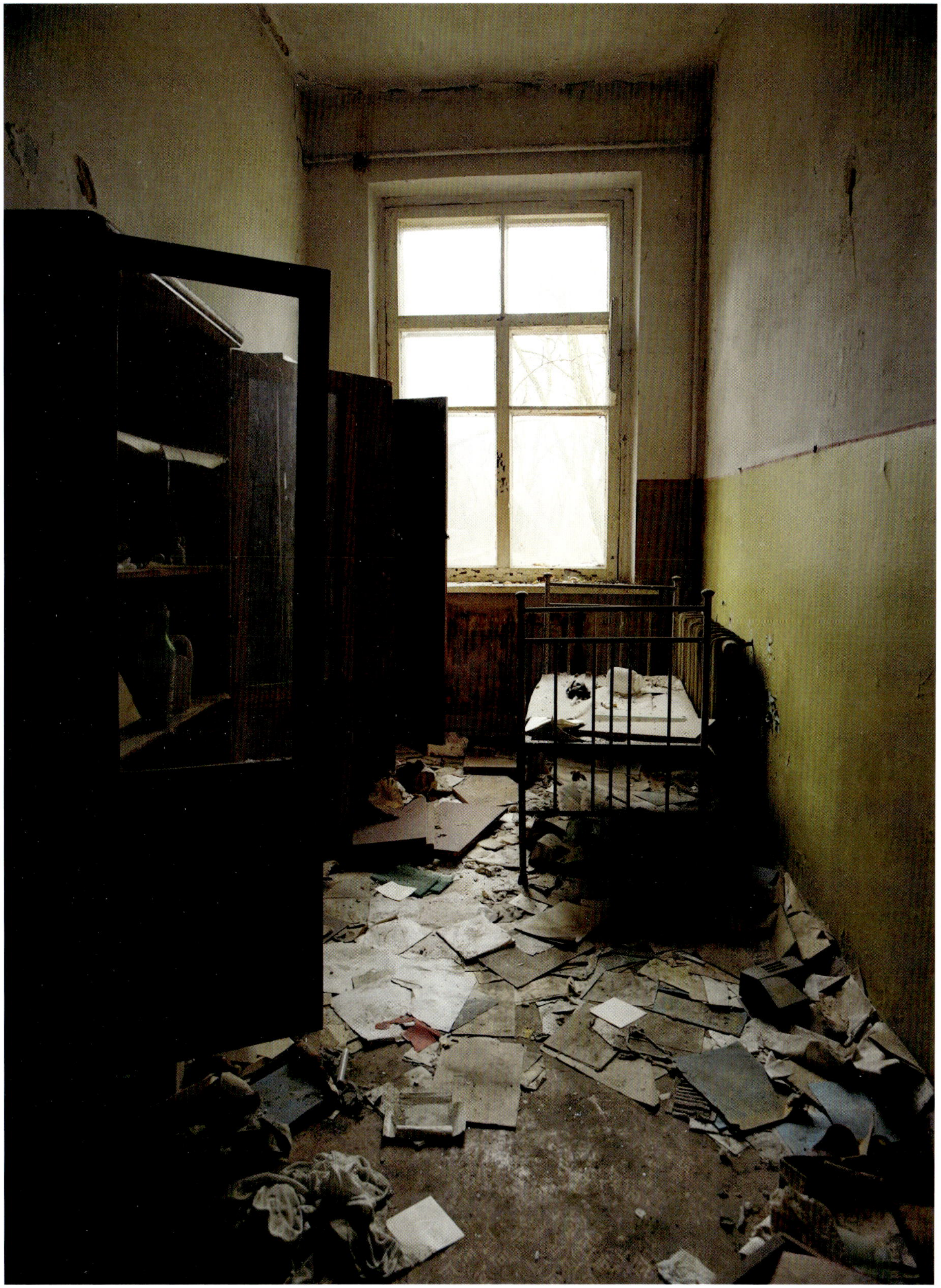

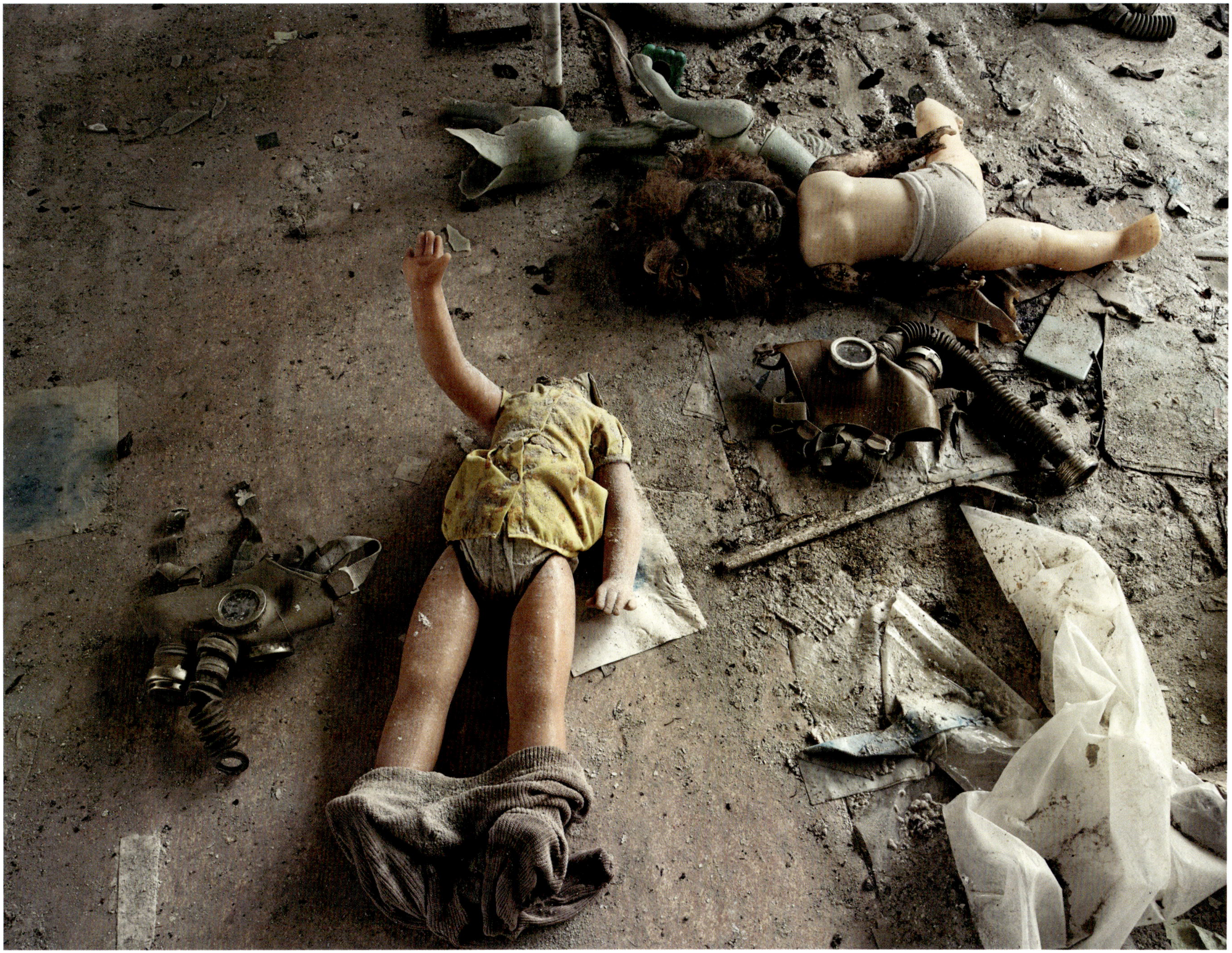

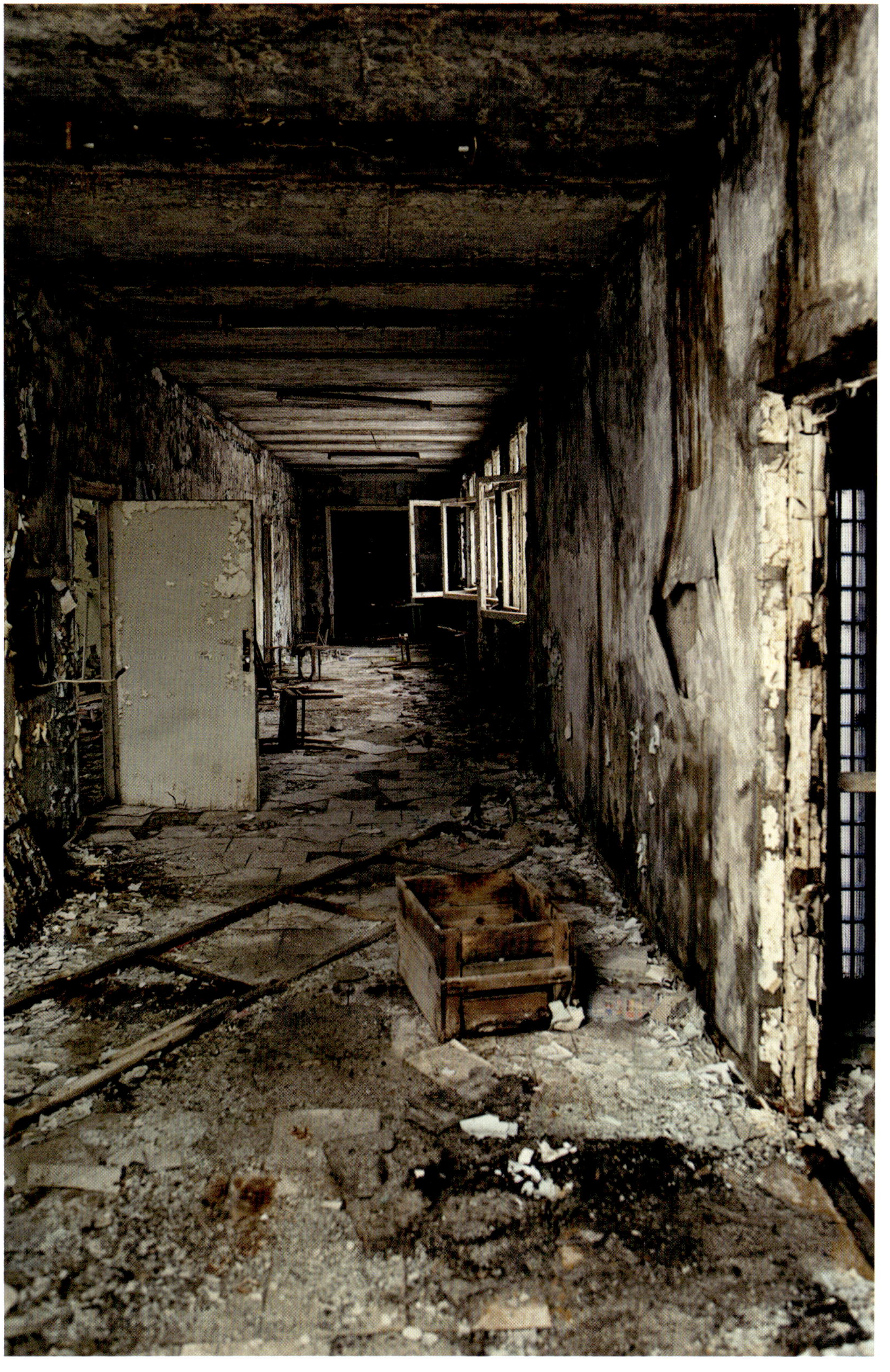

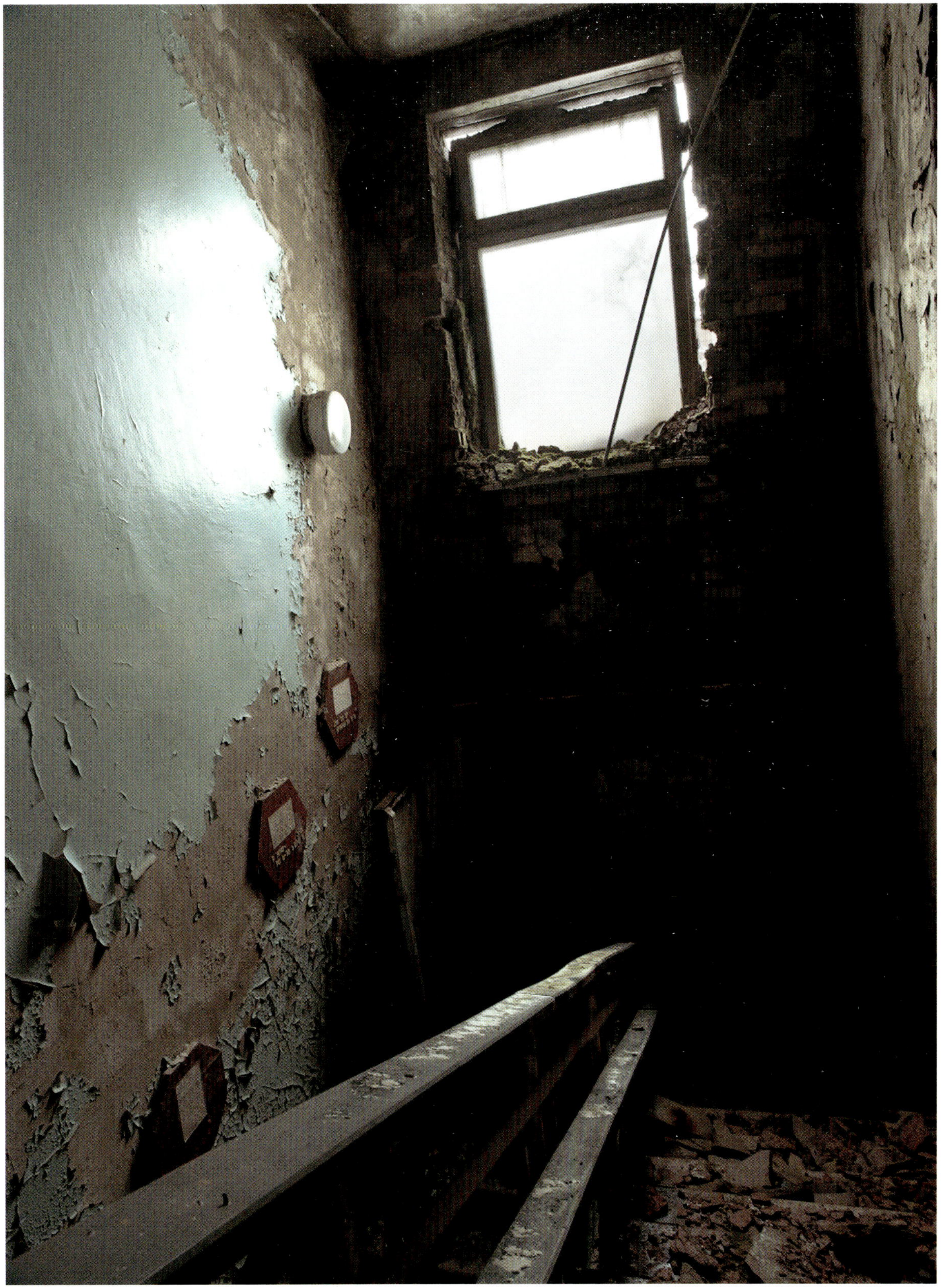

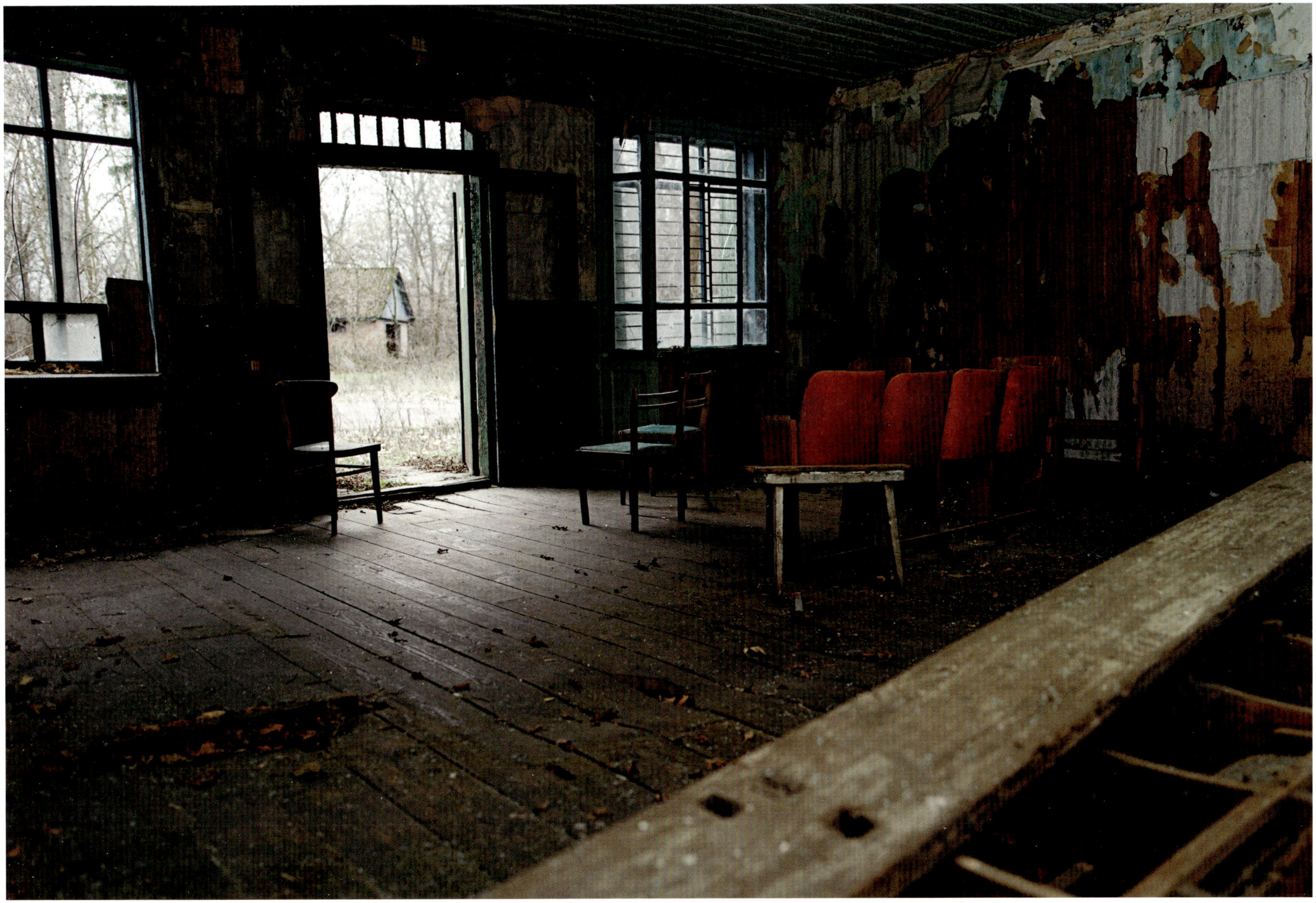

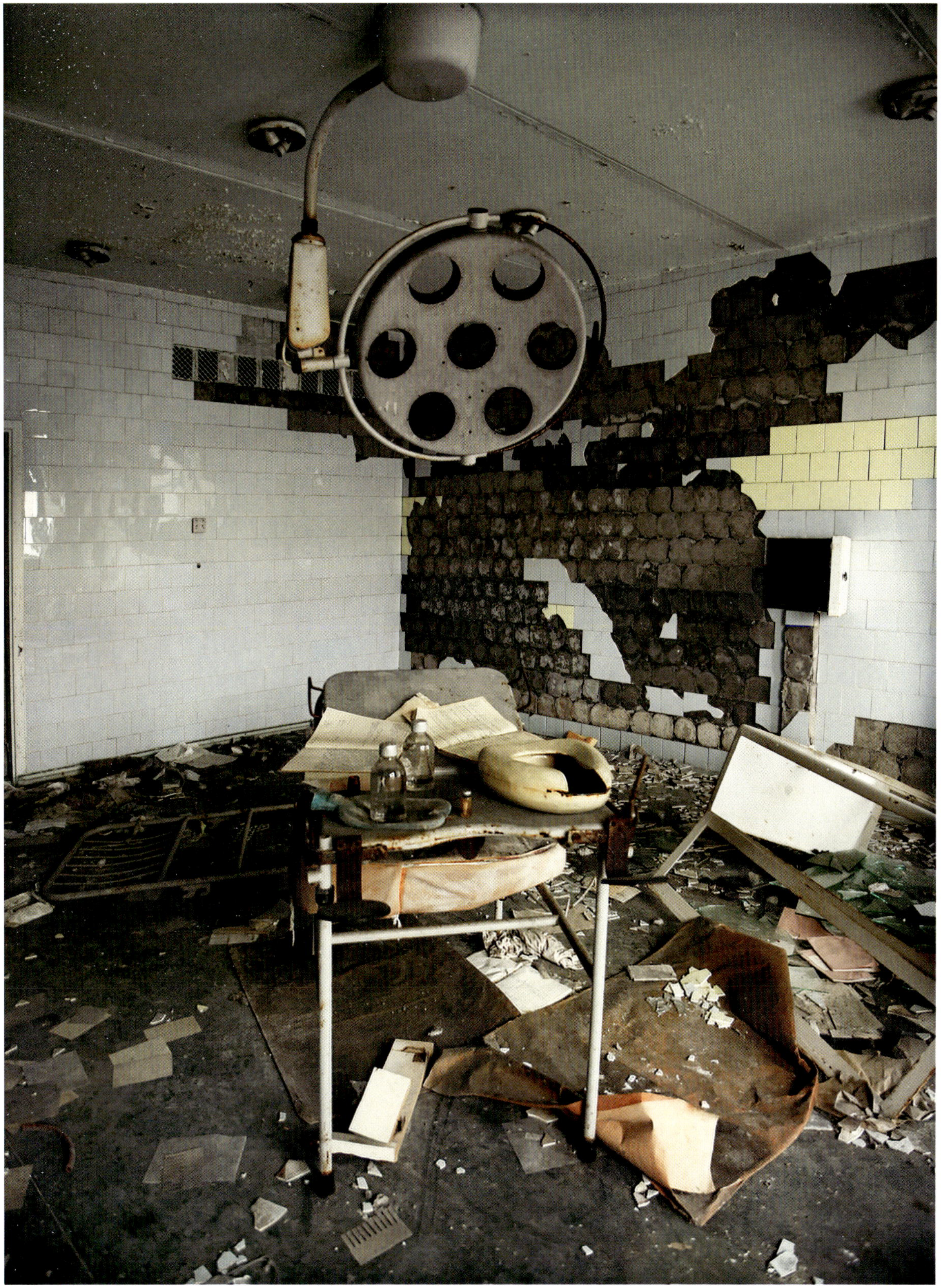

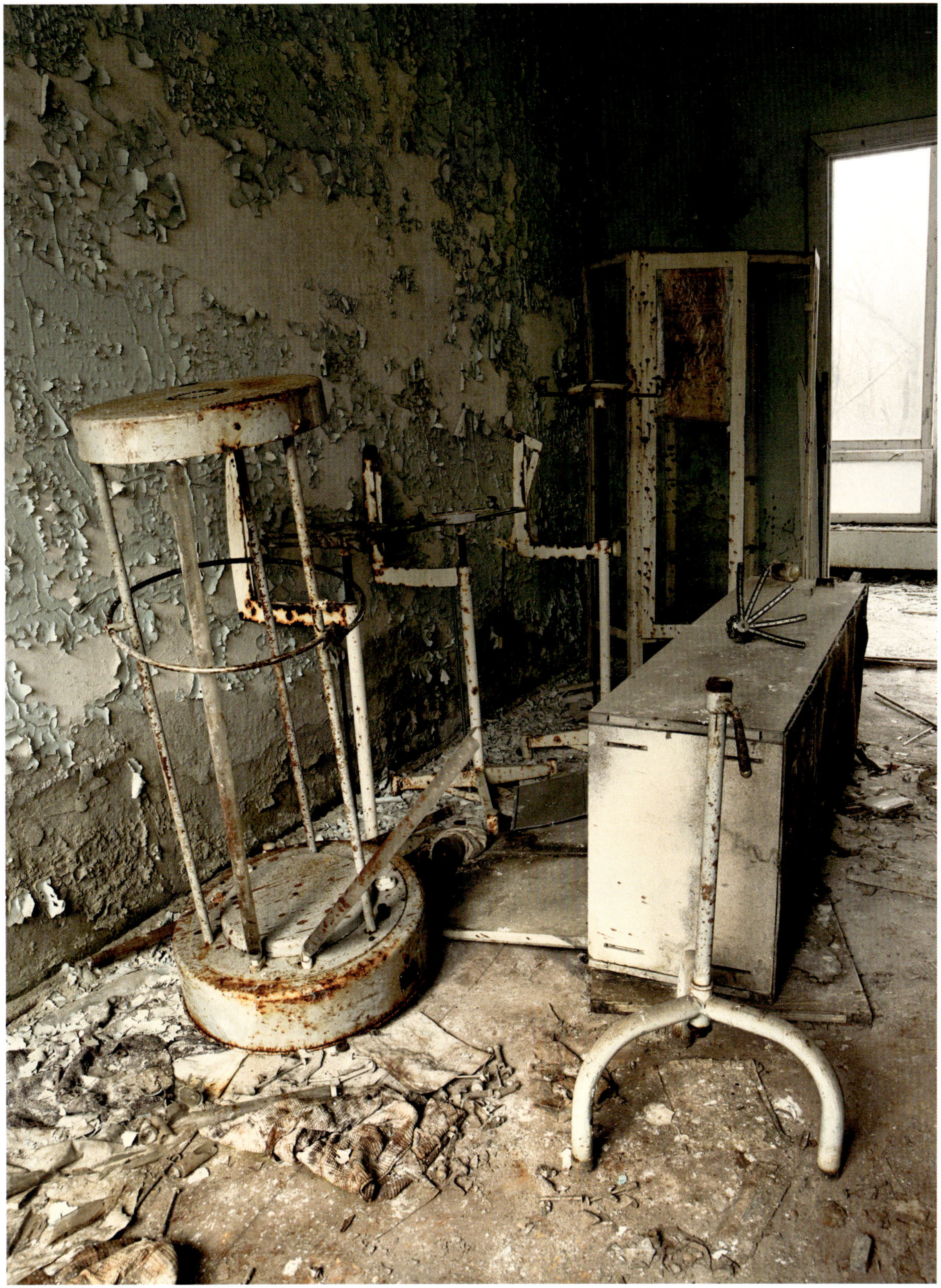

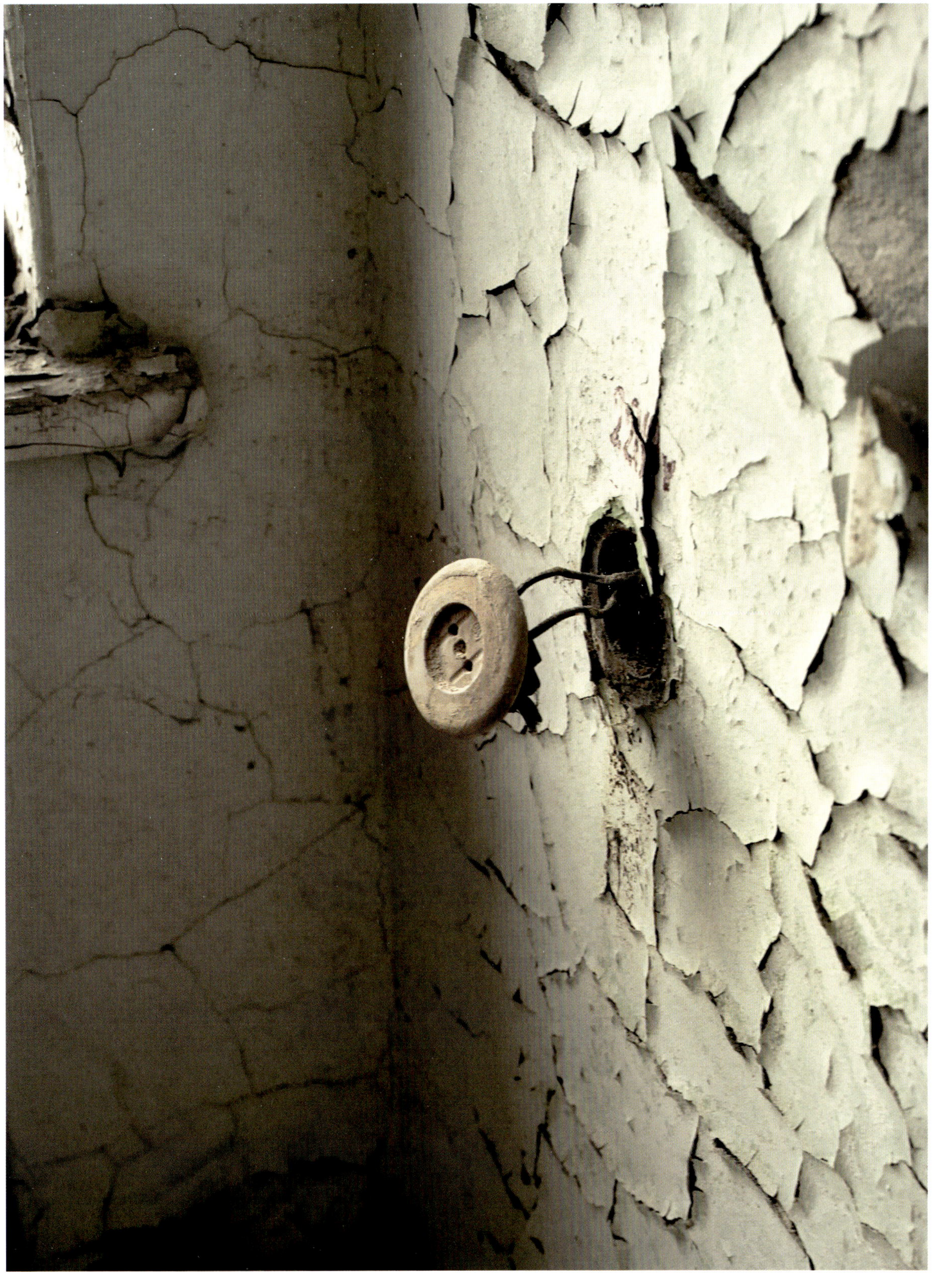

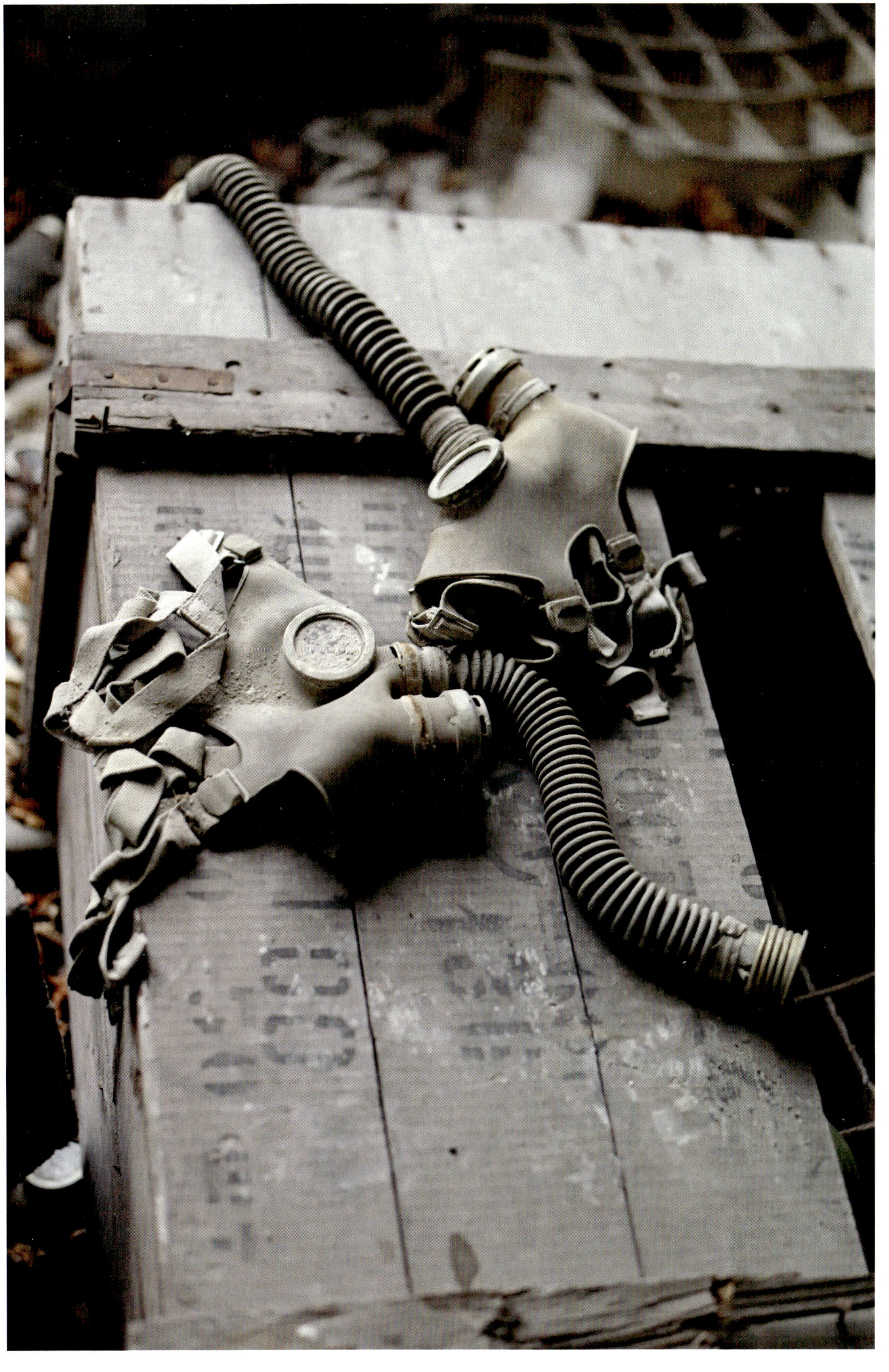

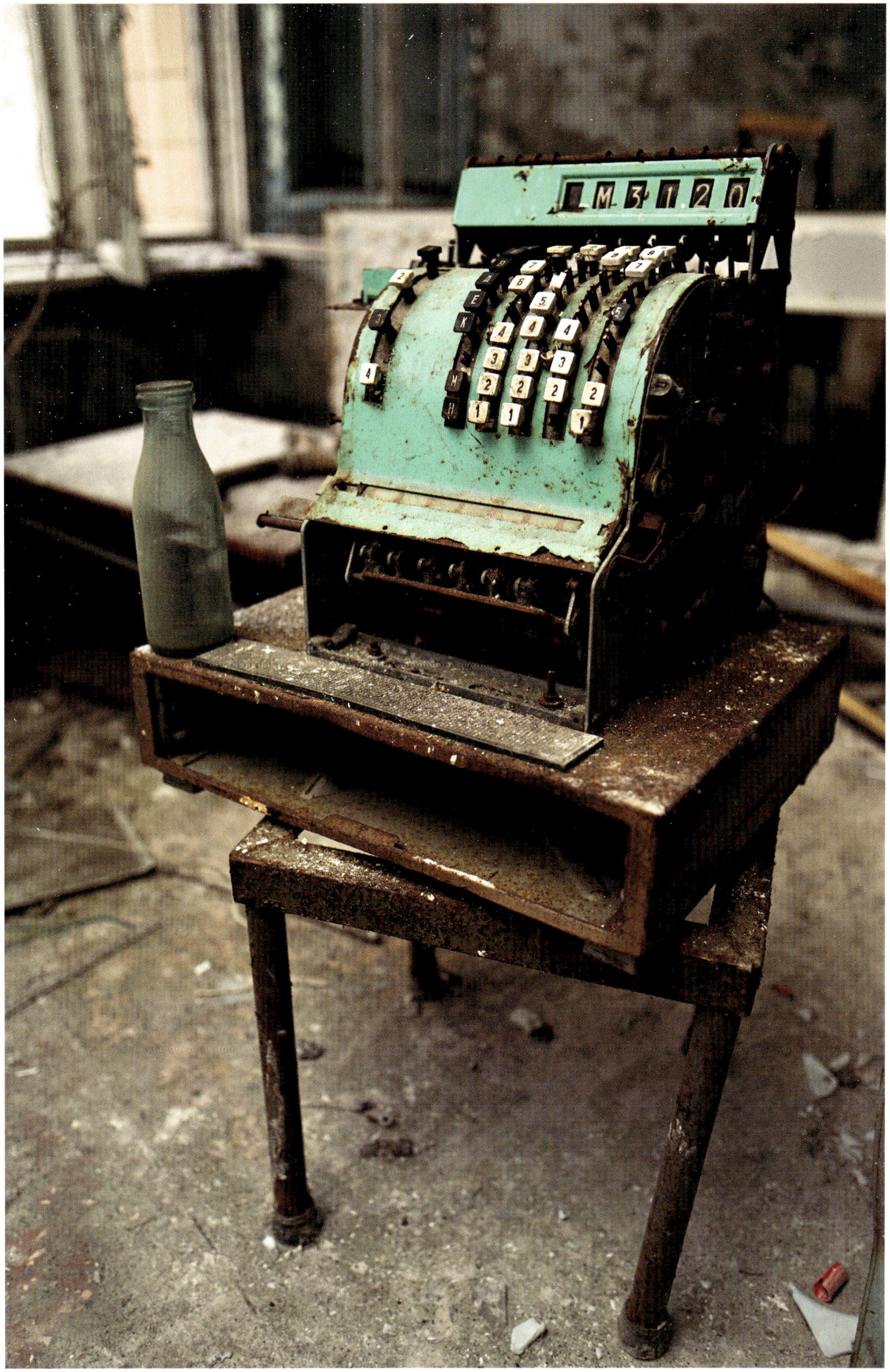
M3120

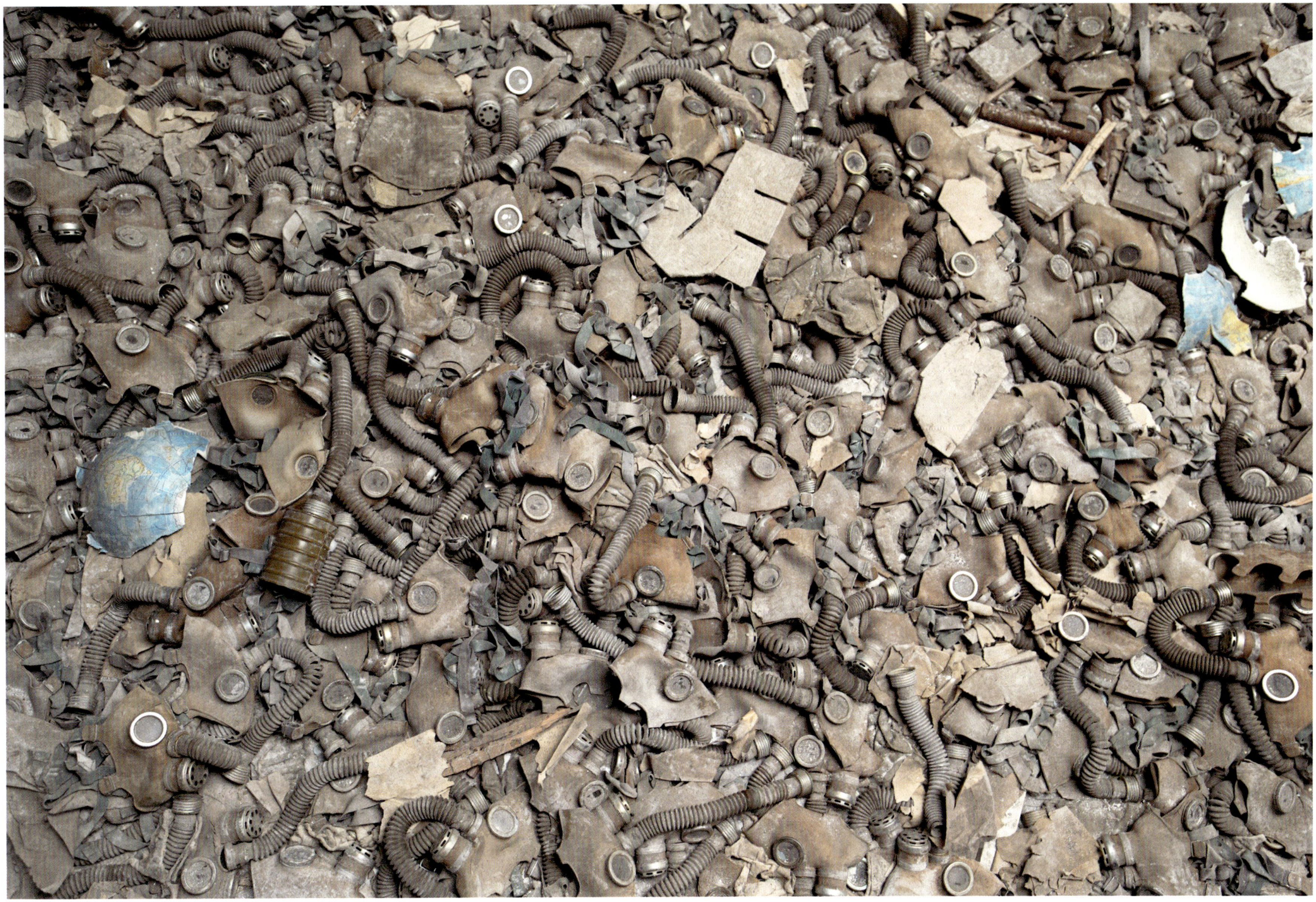

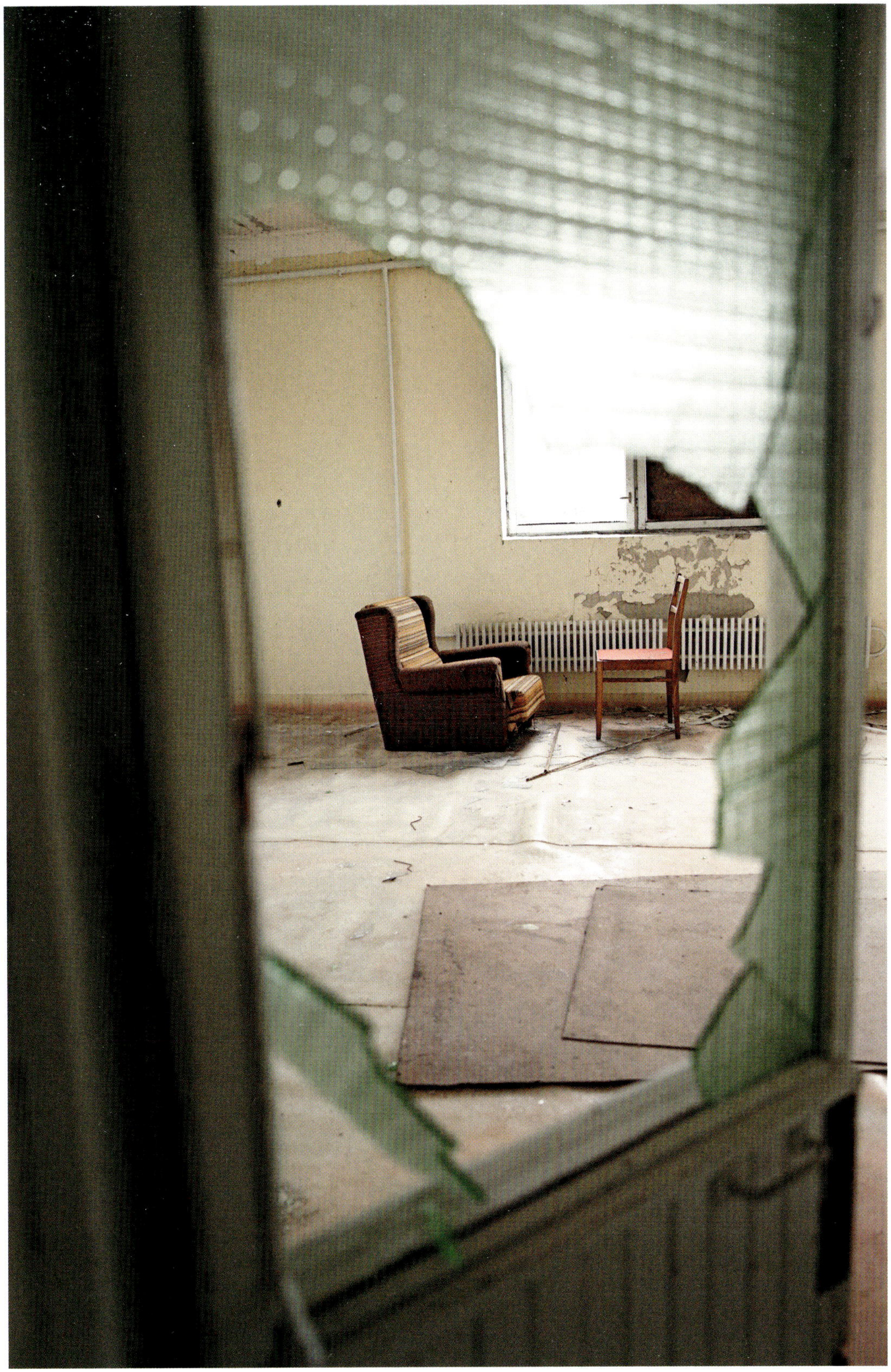

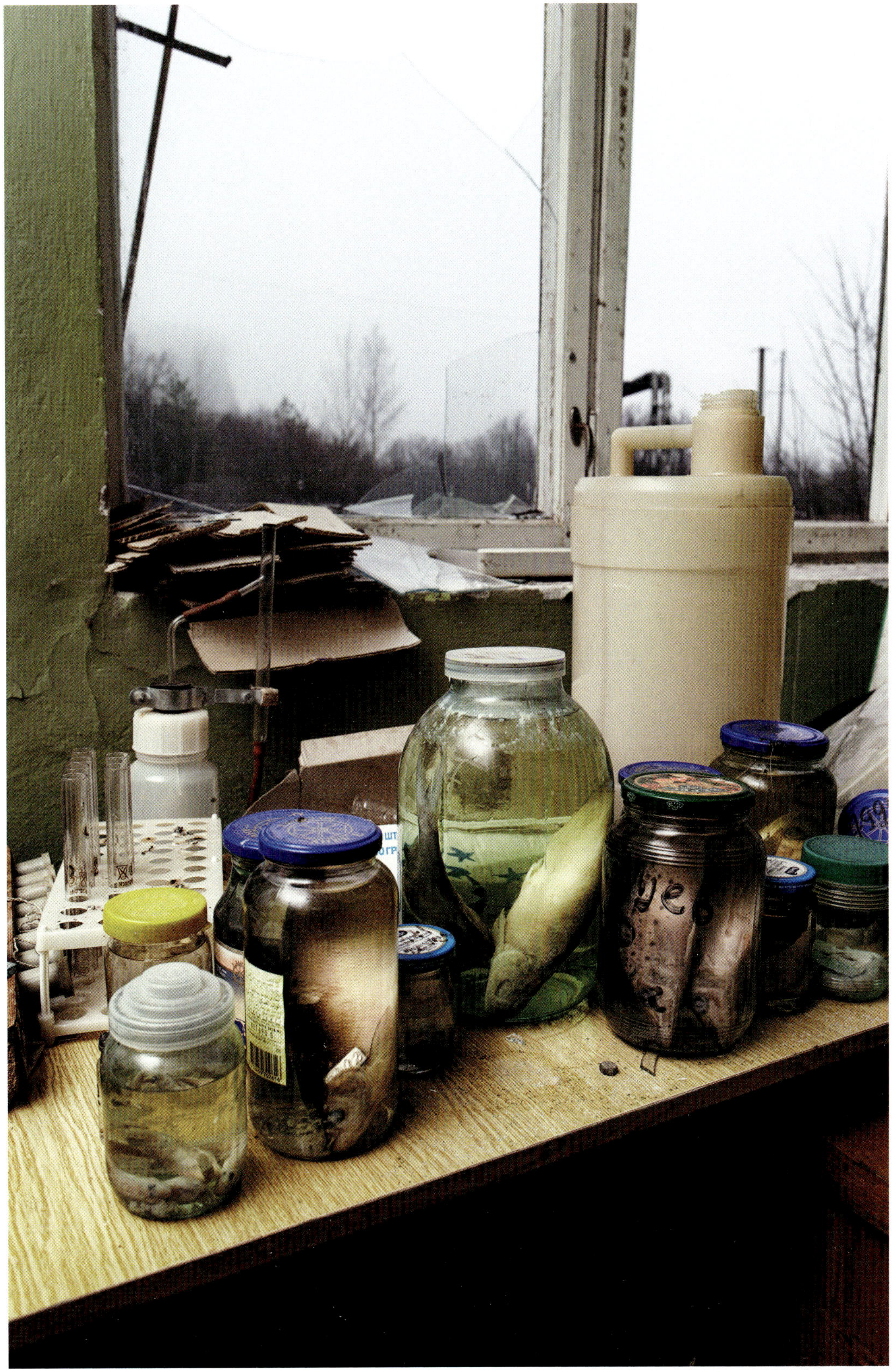

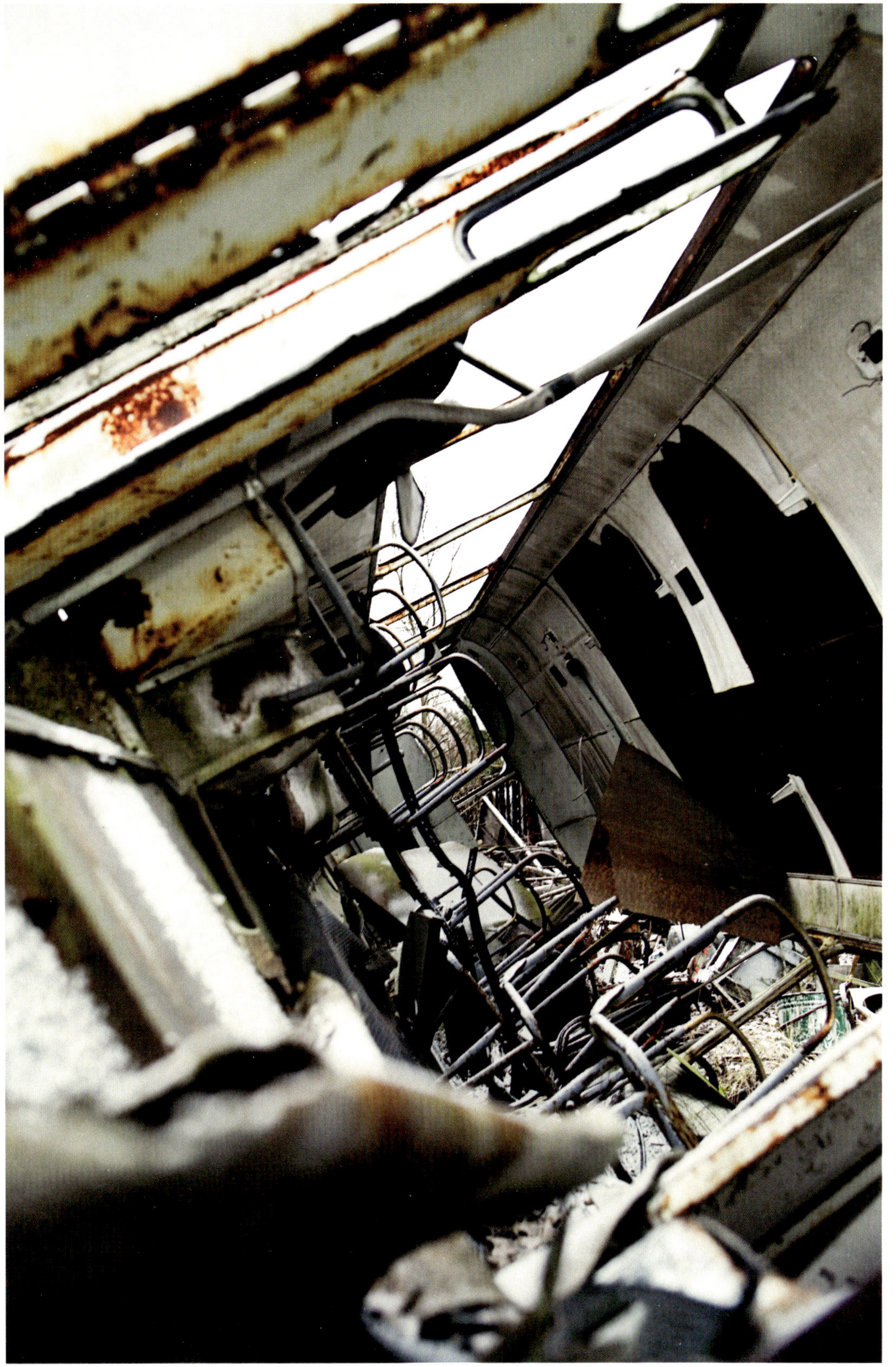

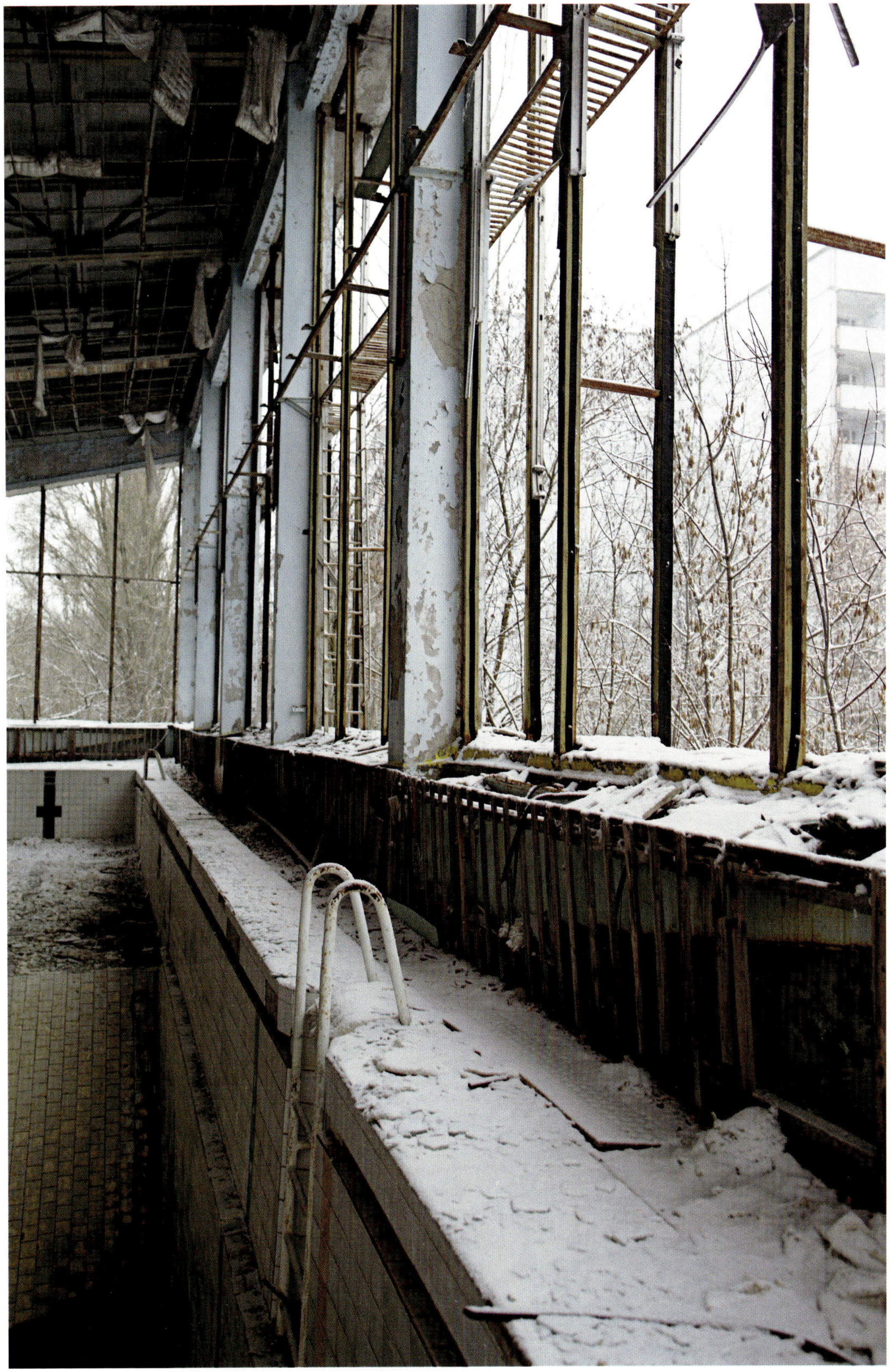

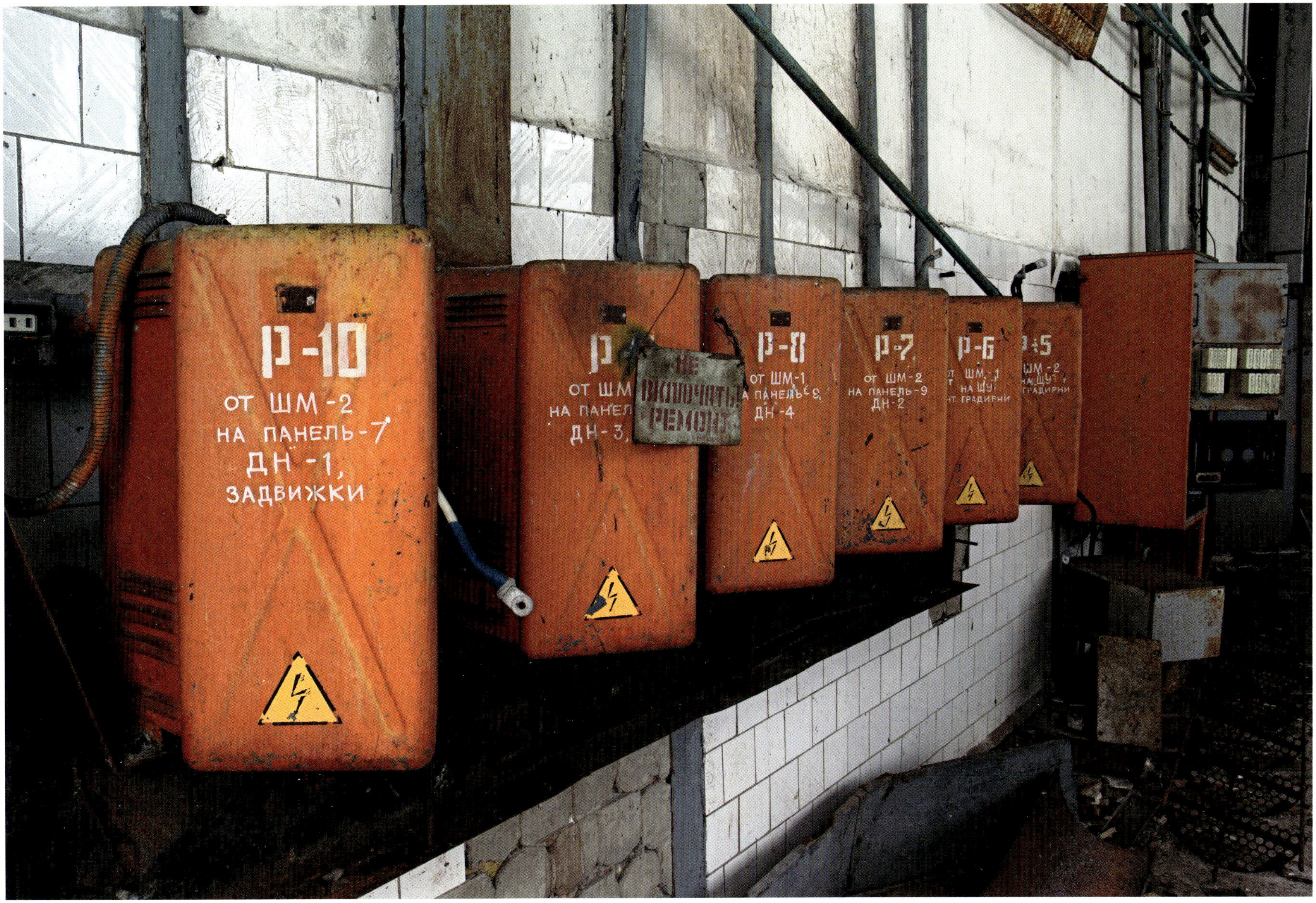
Р-10
ОТ ШМ-2
НА ПАНЕЛЬ-7
ДН-1,
ЗАДВИЖКИ
РЕМОНТ
Р-8
ОТ ШМ-1
ДН-4
Р-7
ОТ ШМ-2
НА ПАНЕЛЬ-9
ДН-2
Р-6

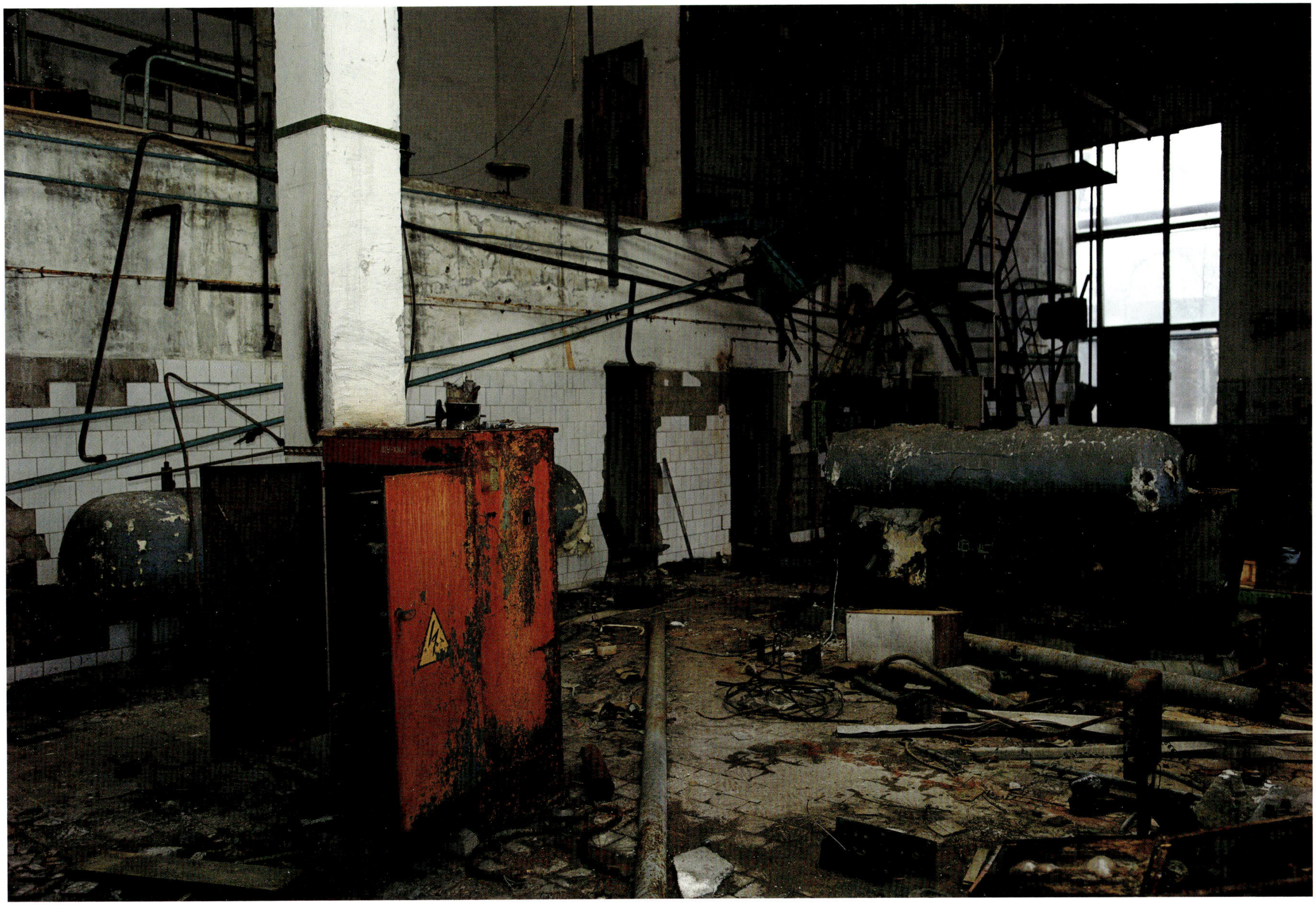

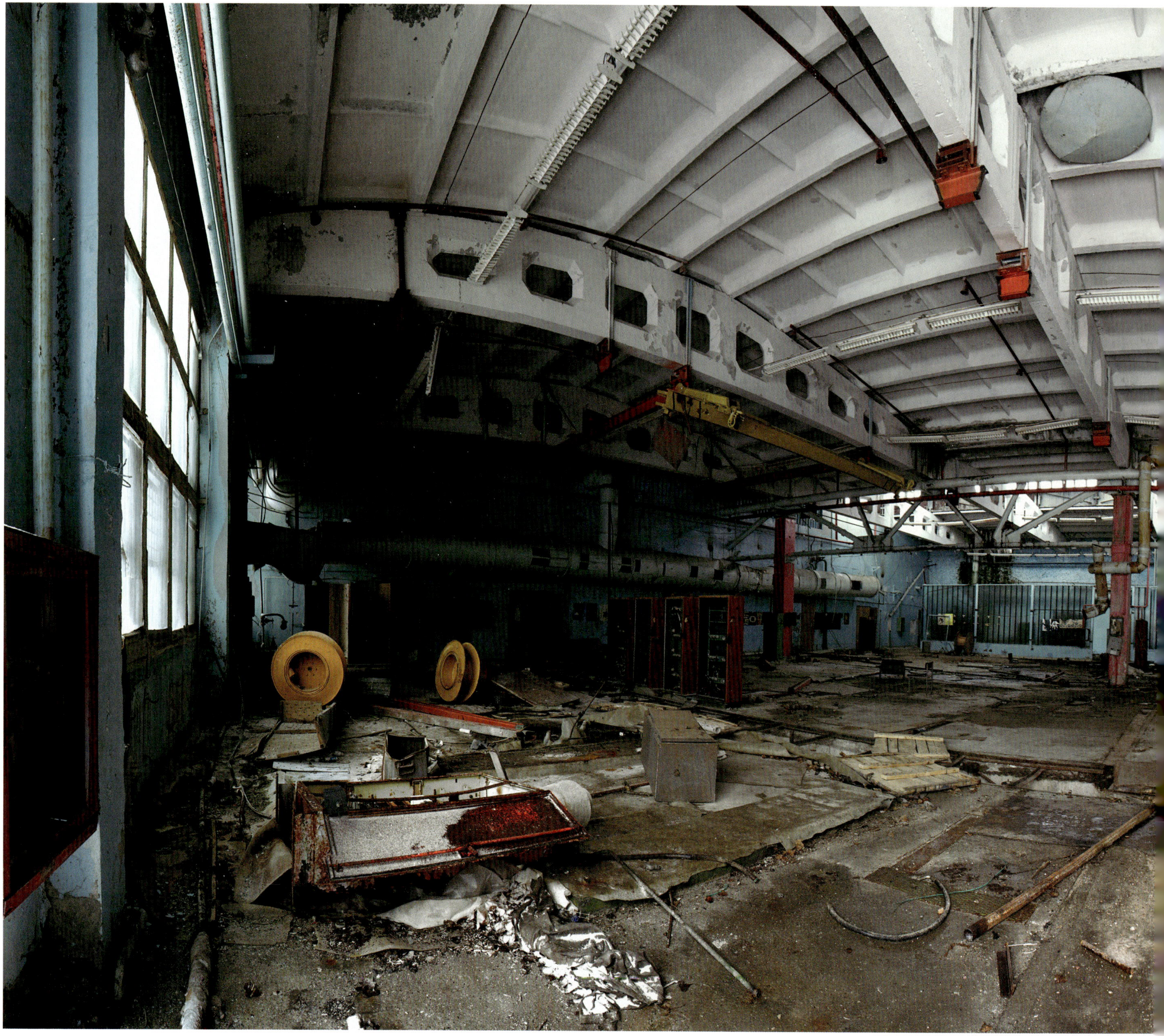

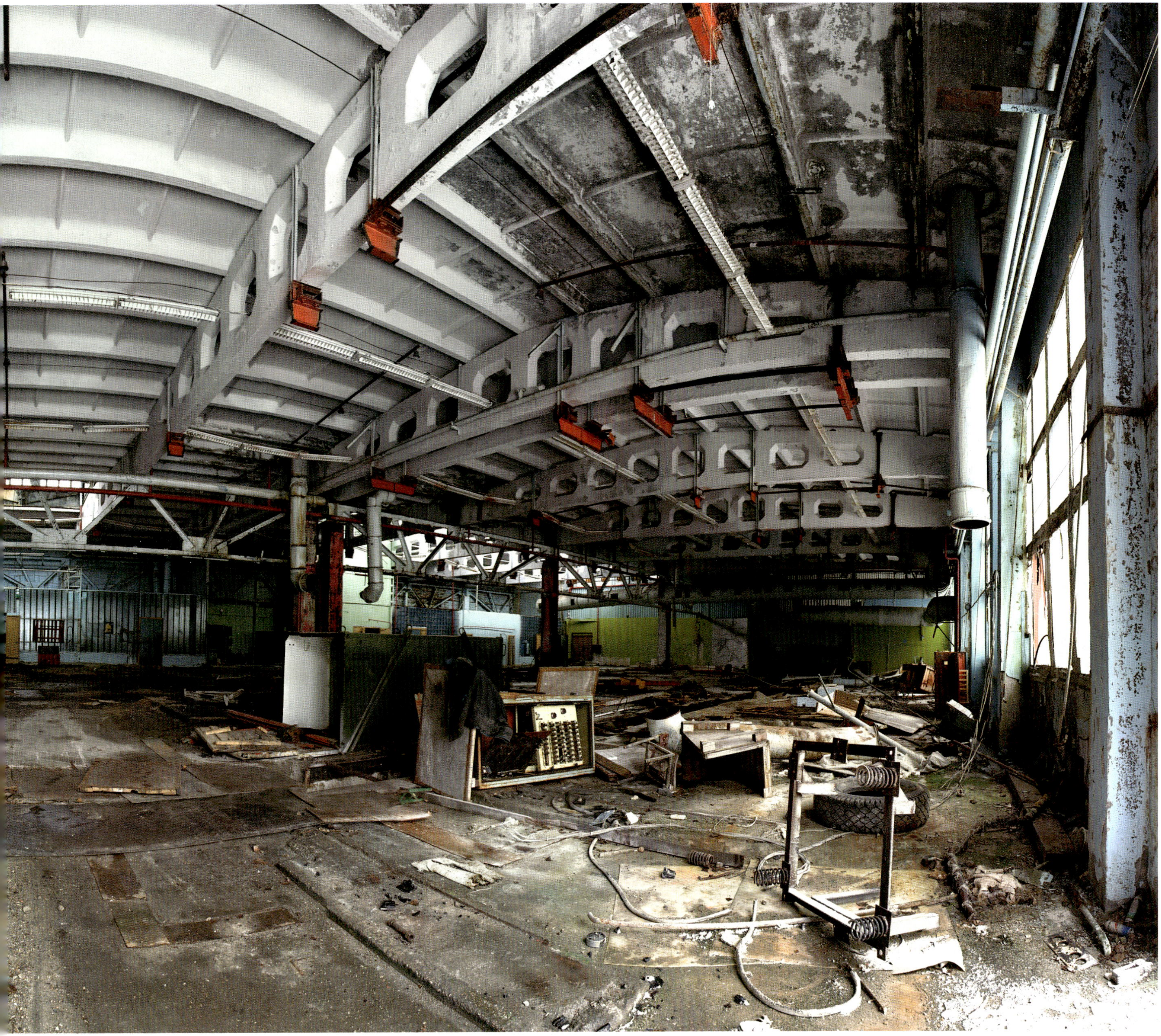

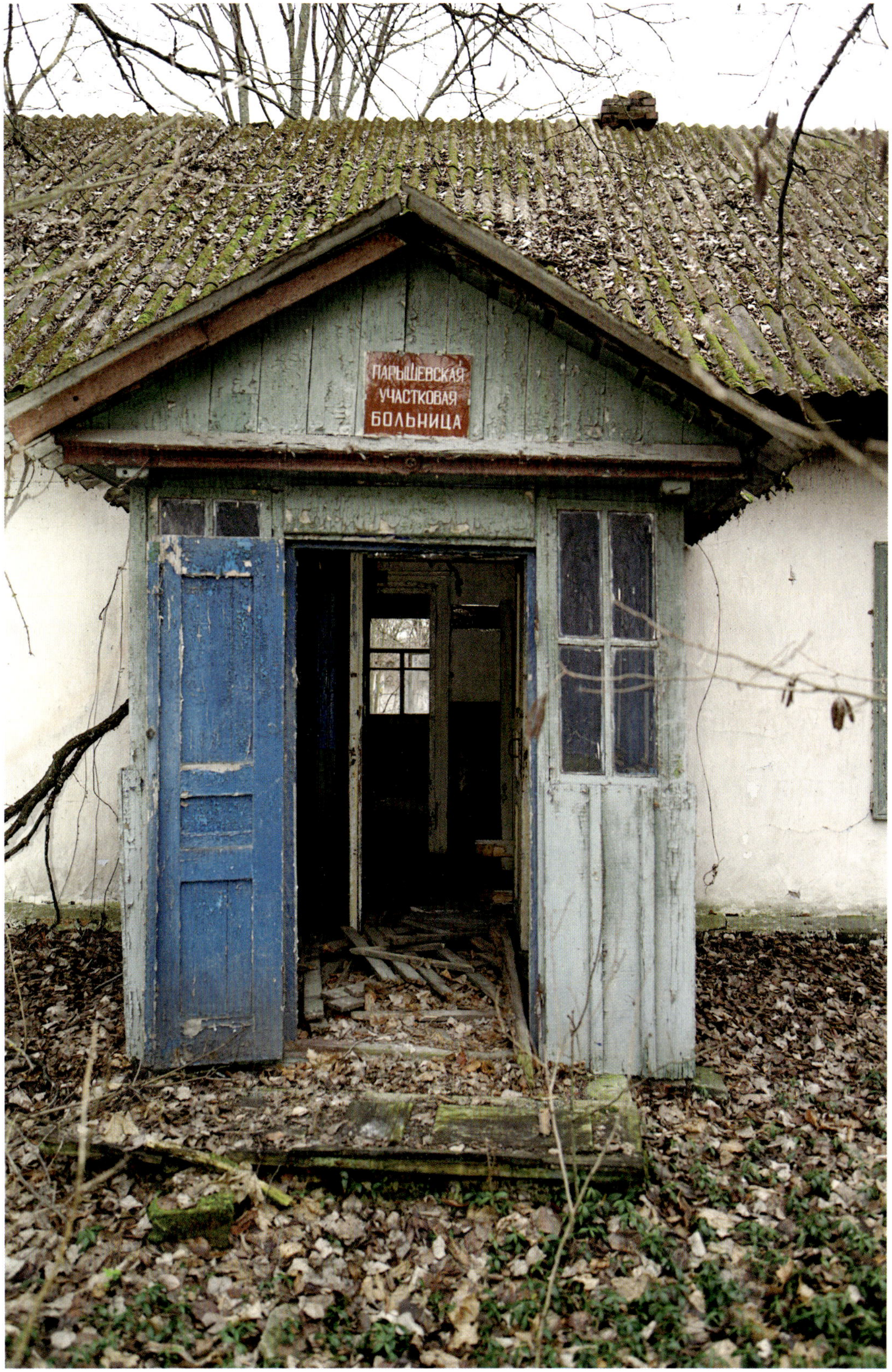
ПАРЫШЕВСКАЯ
УЧАСТКОВАЯ
БОЛЬНИЦА

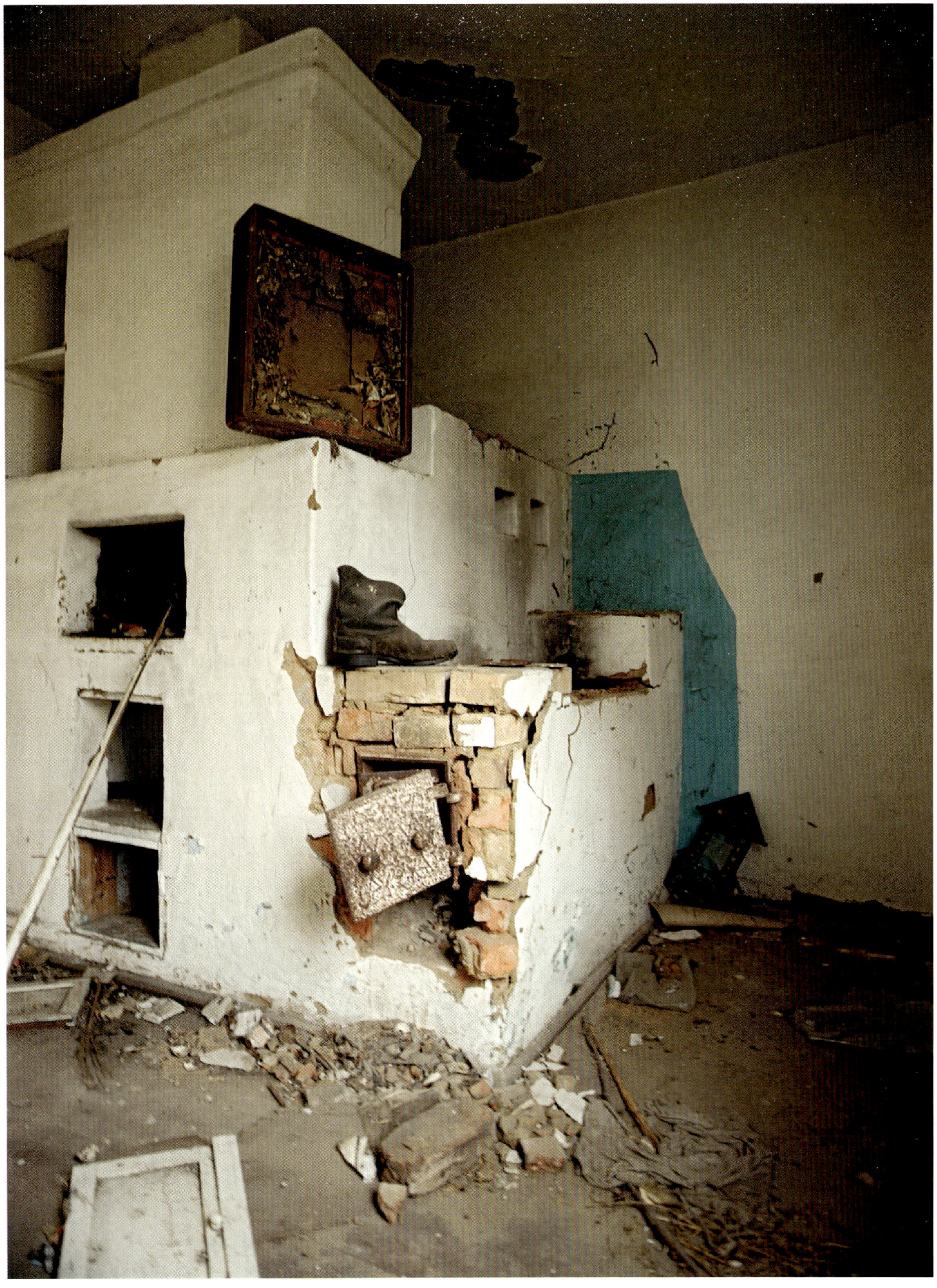

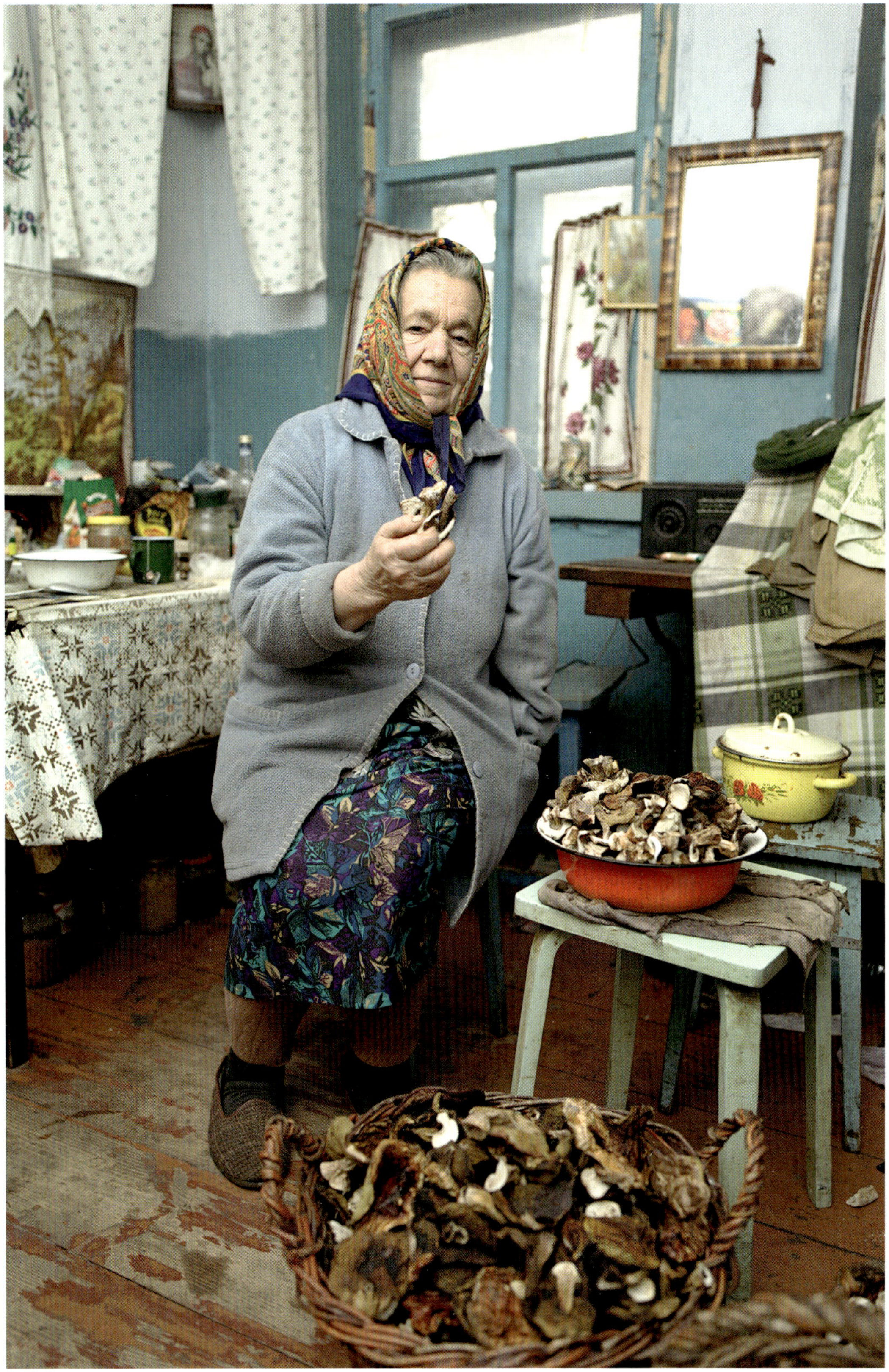

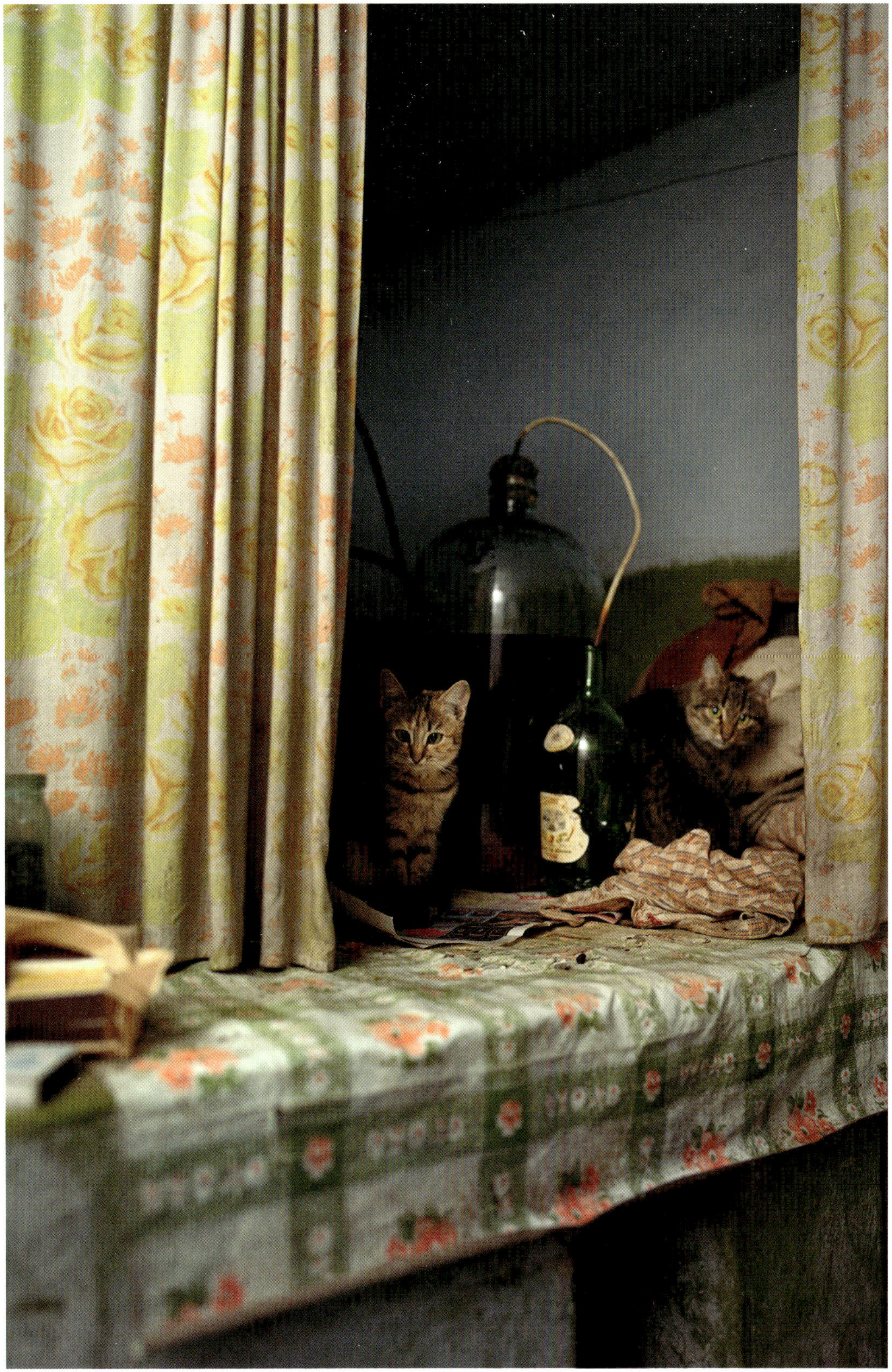

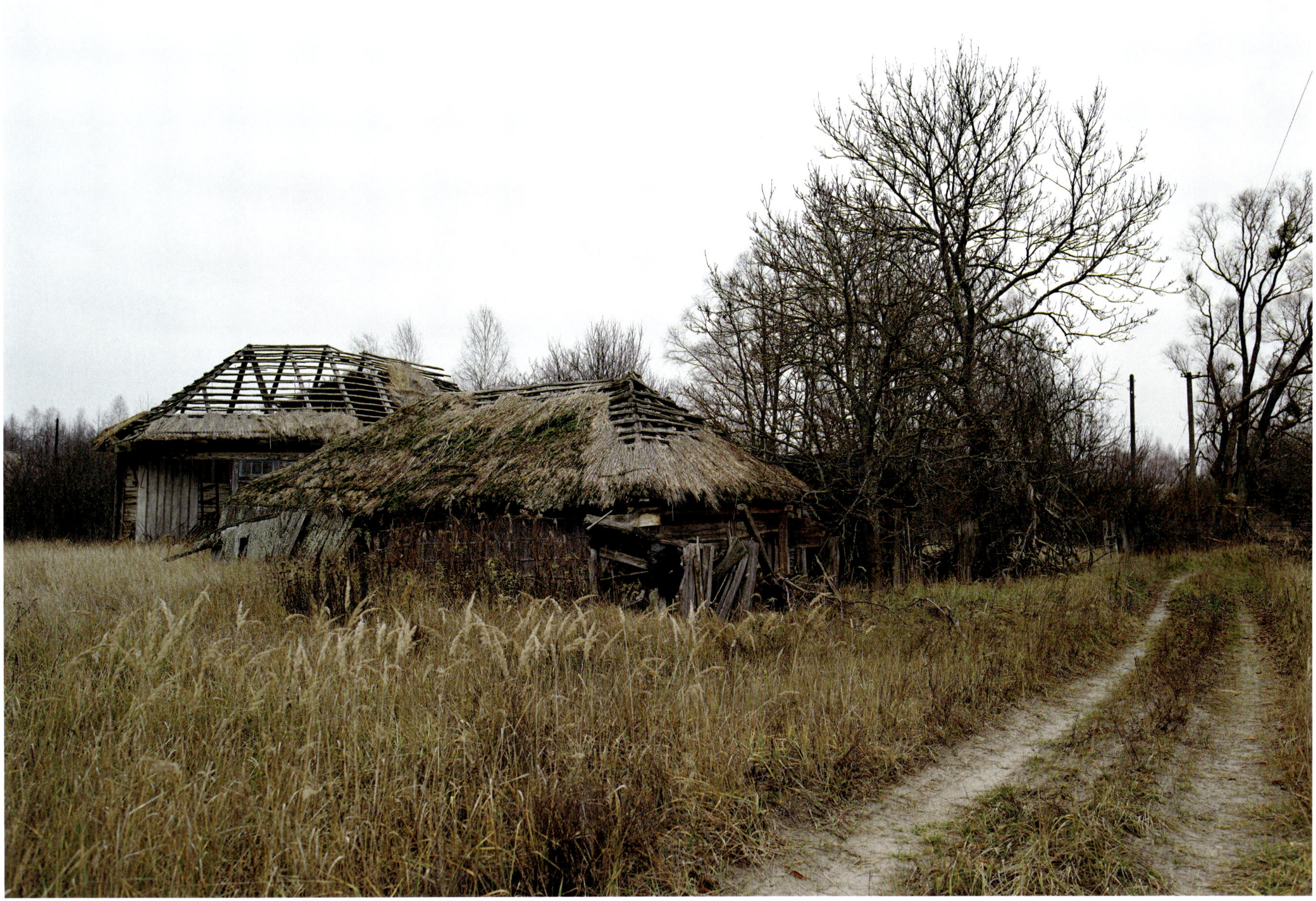

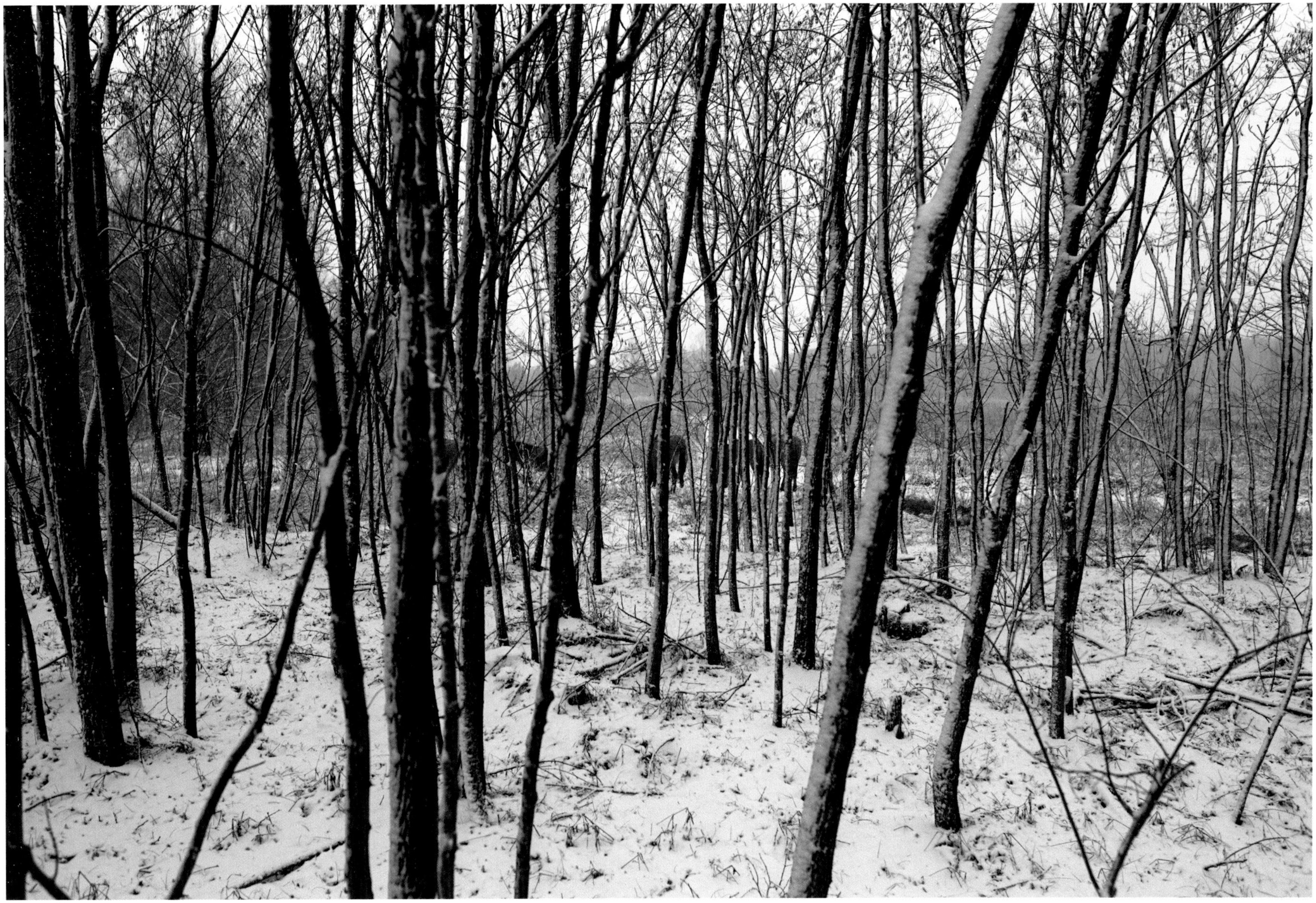

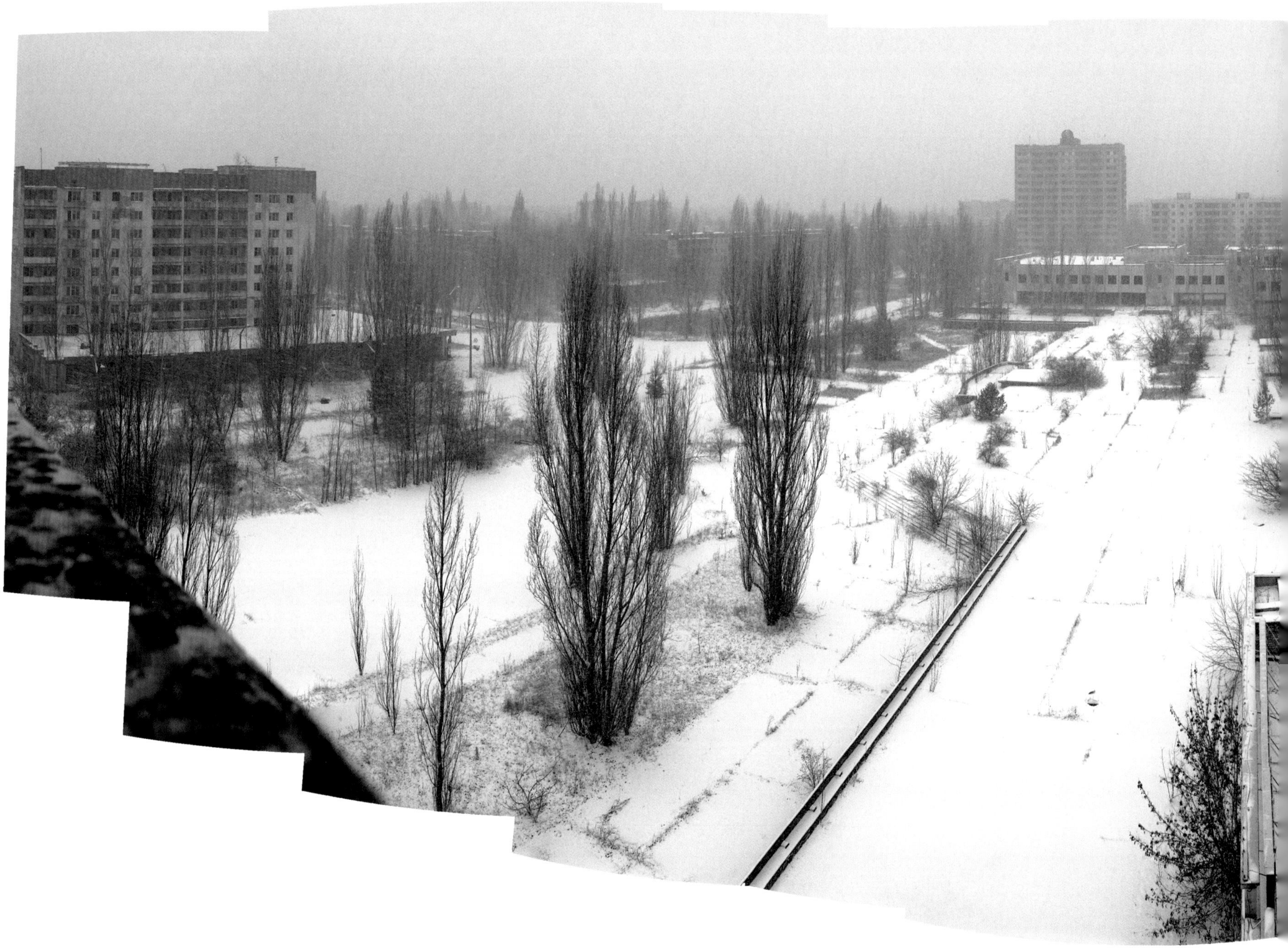

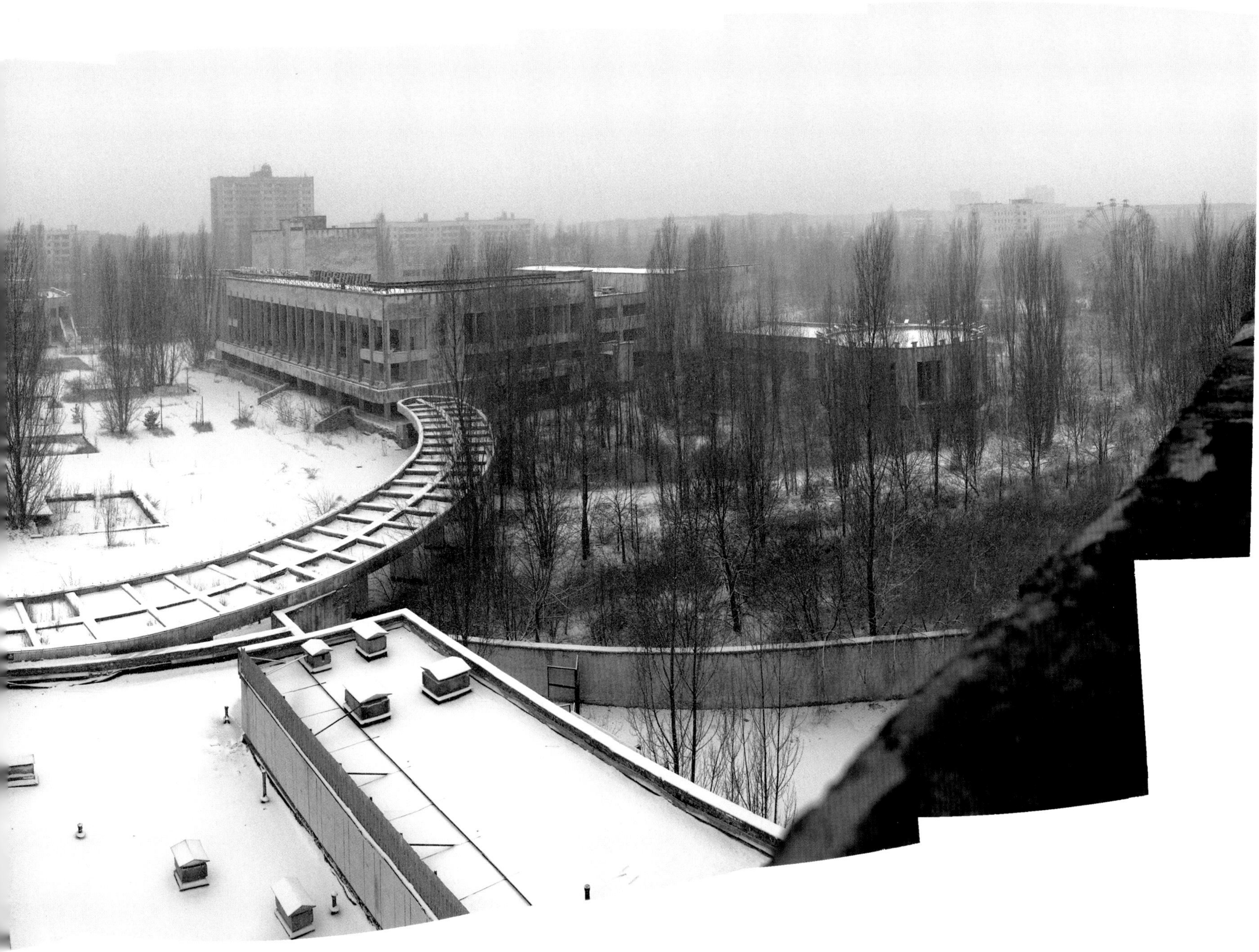

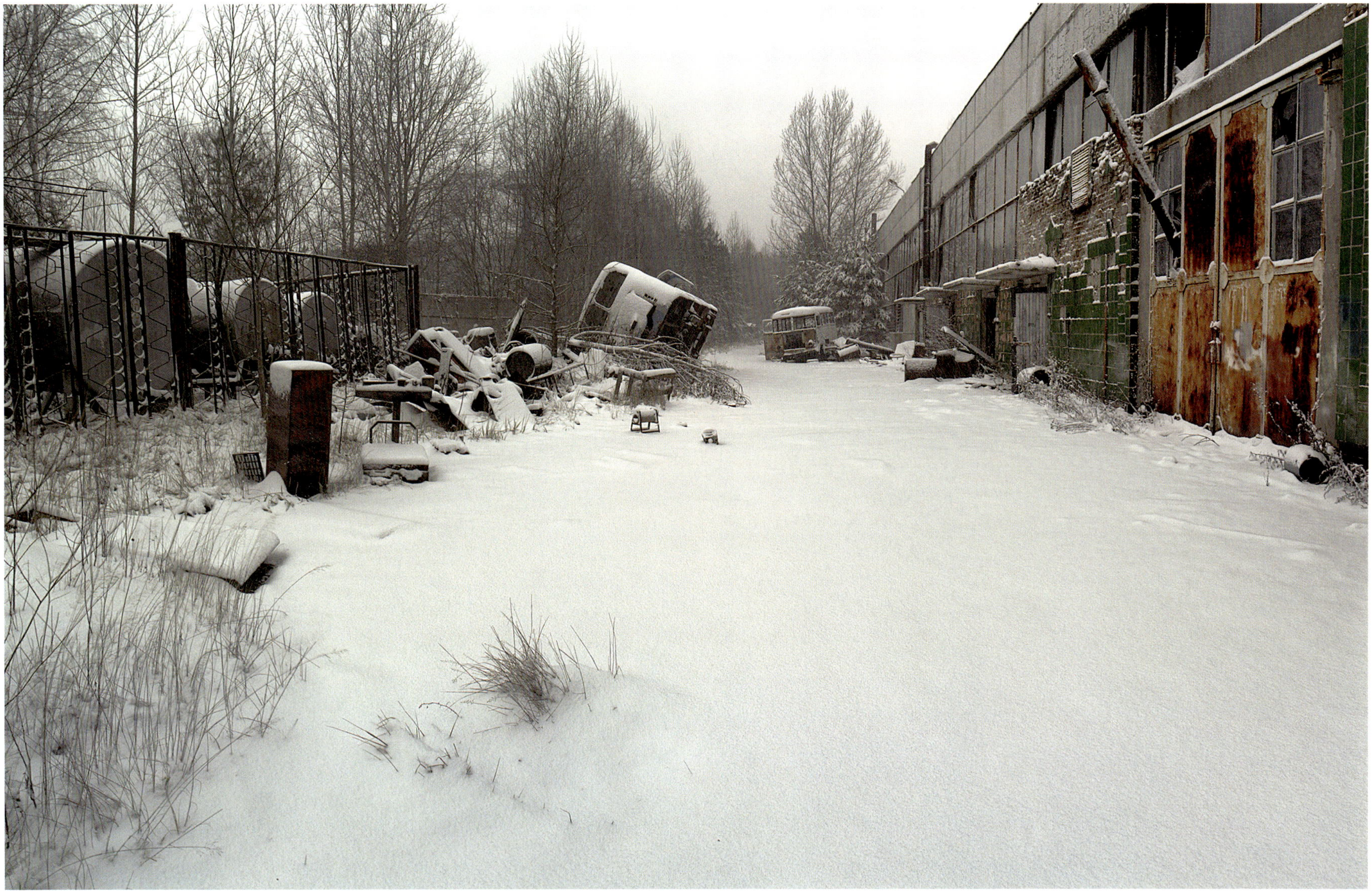

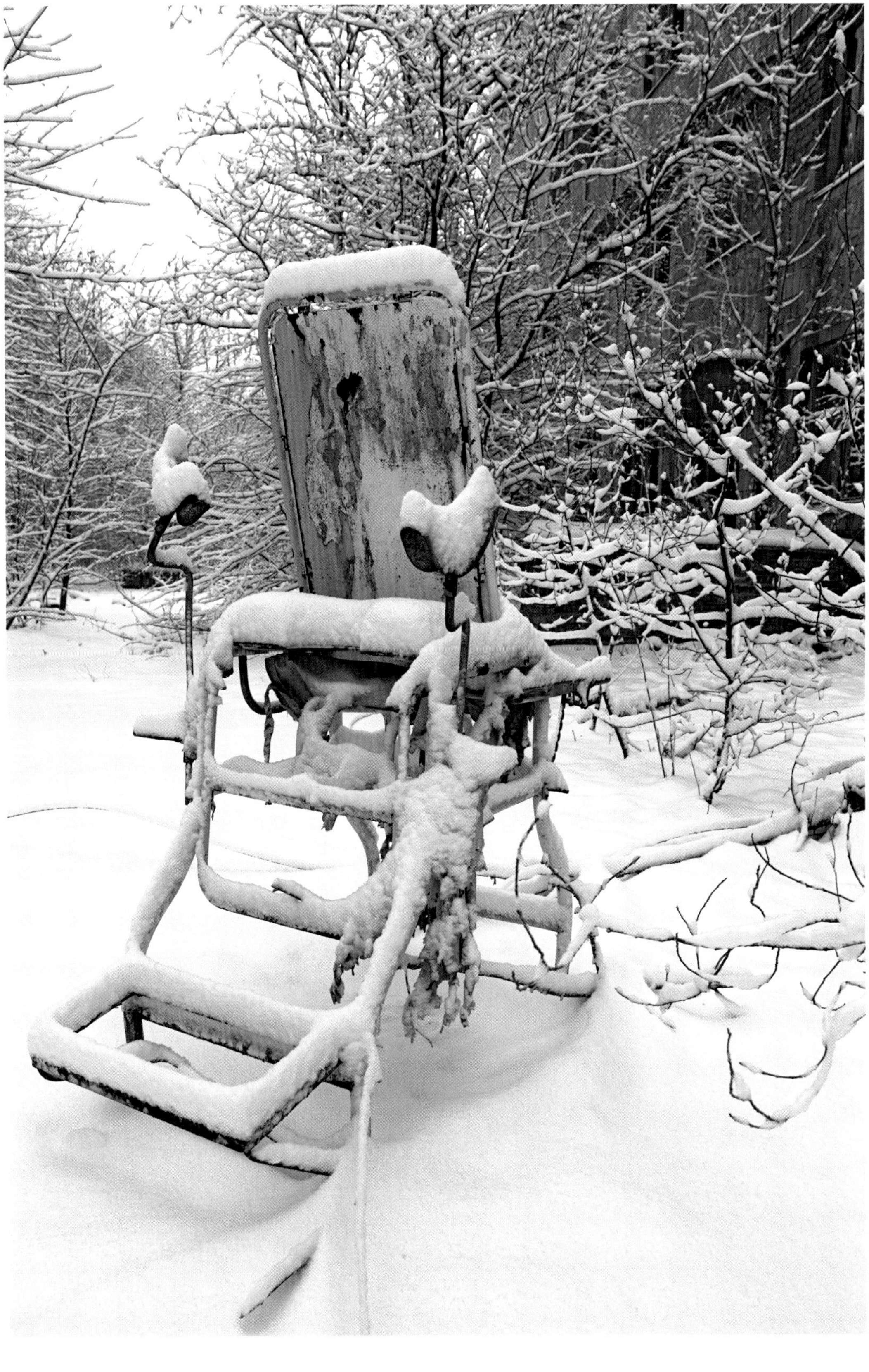

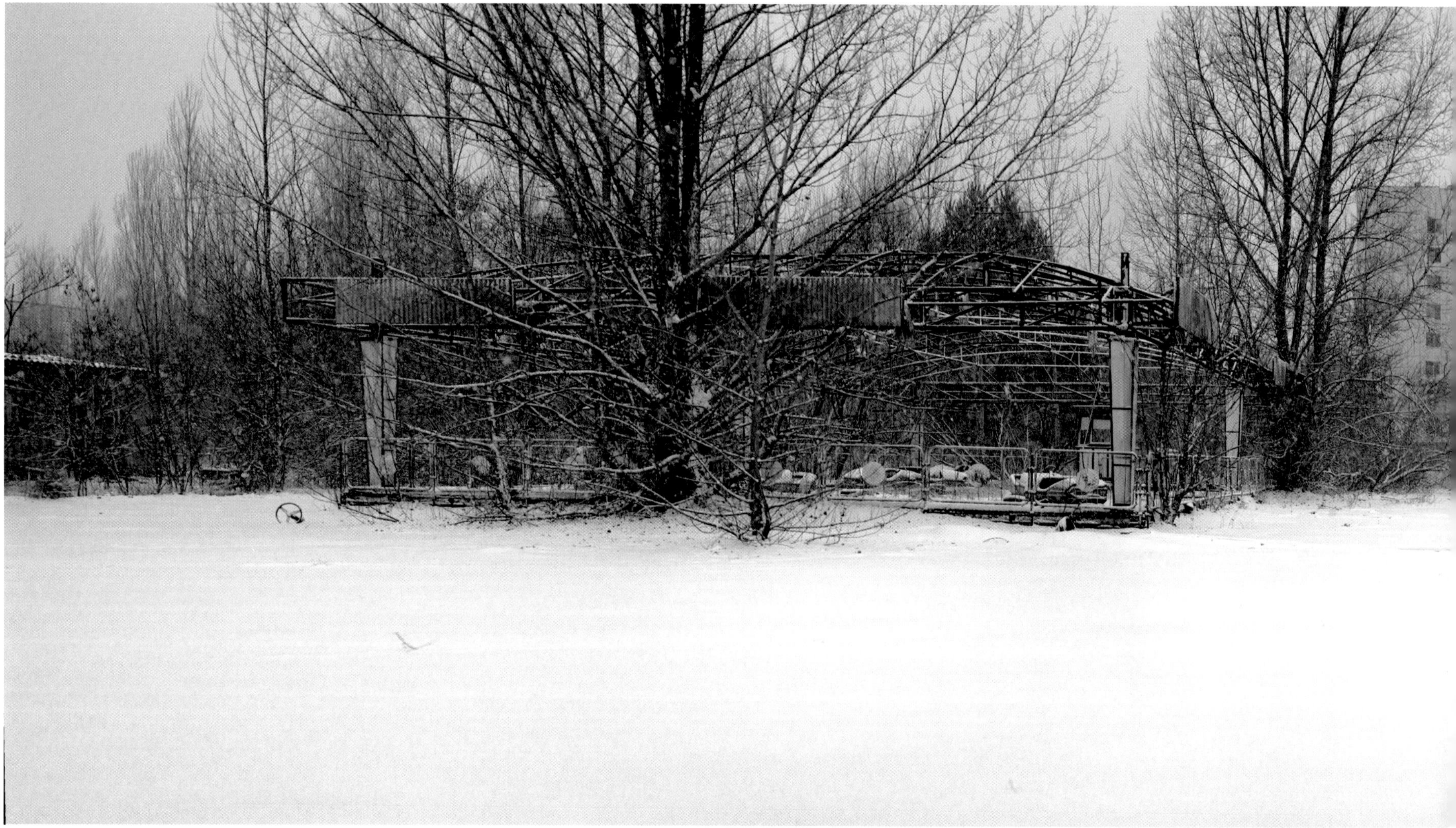

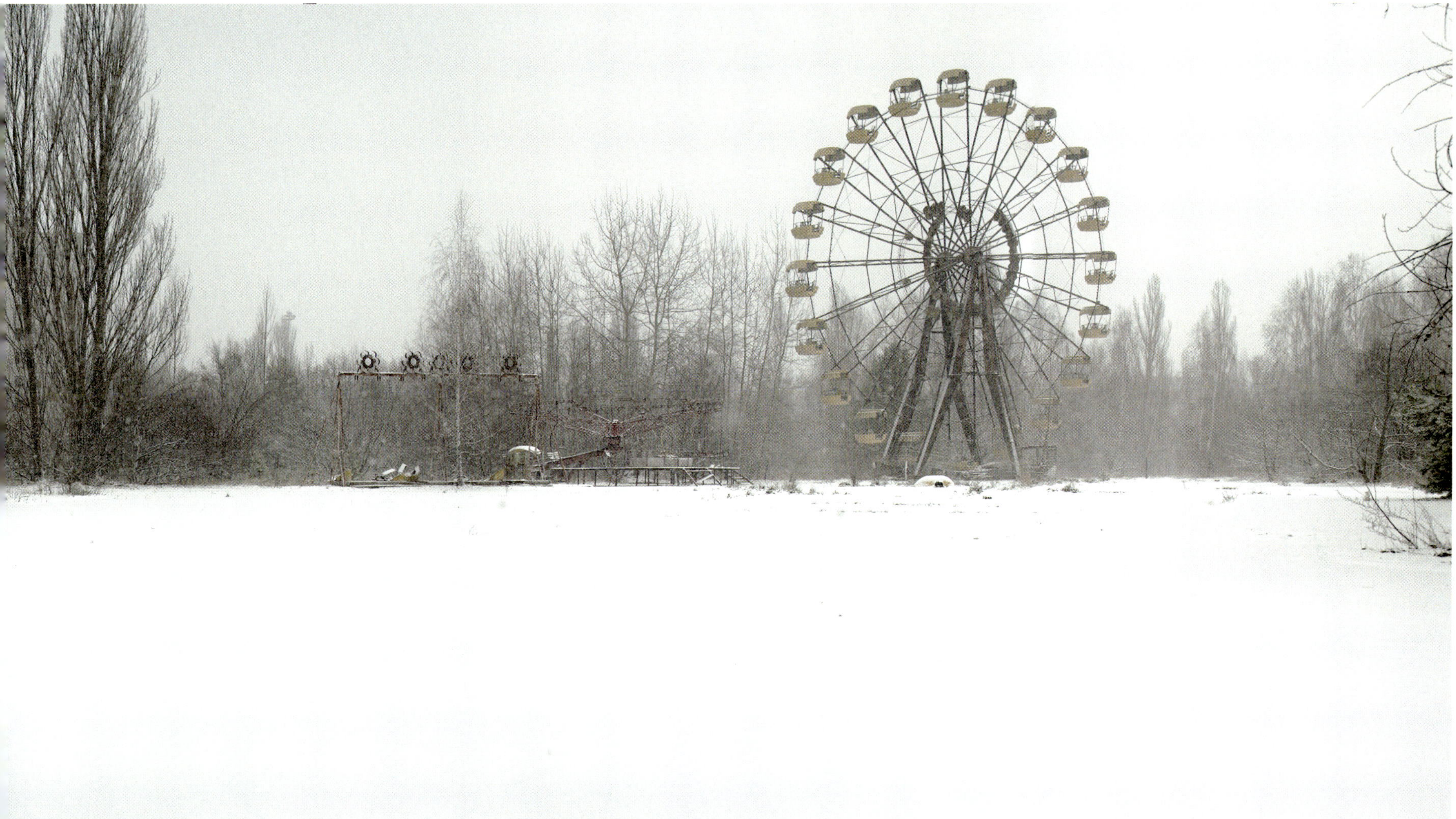

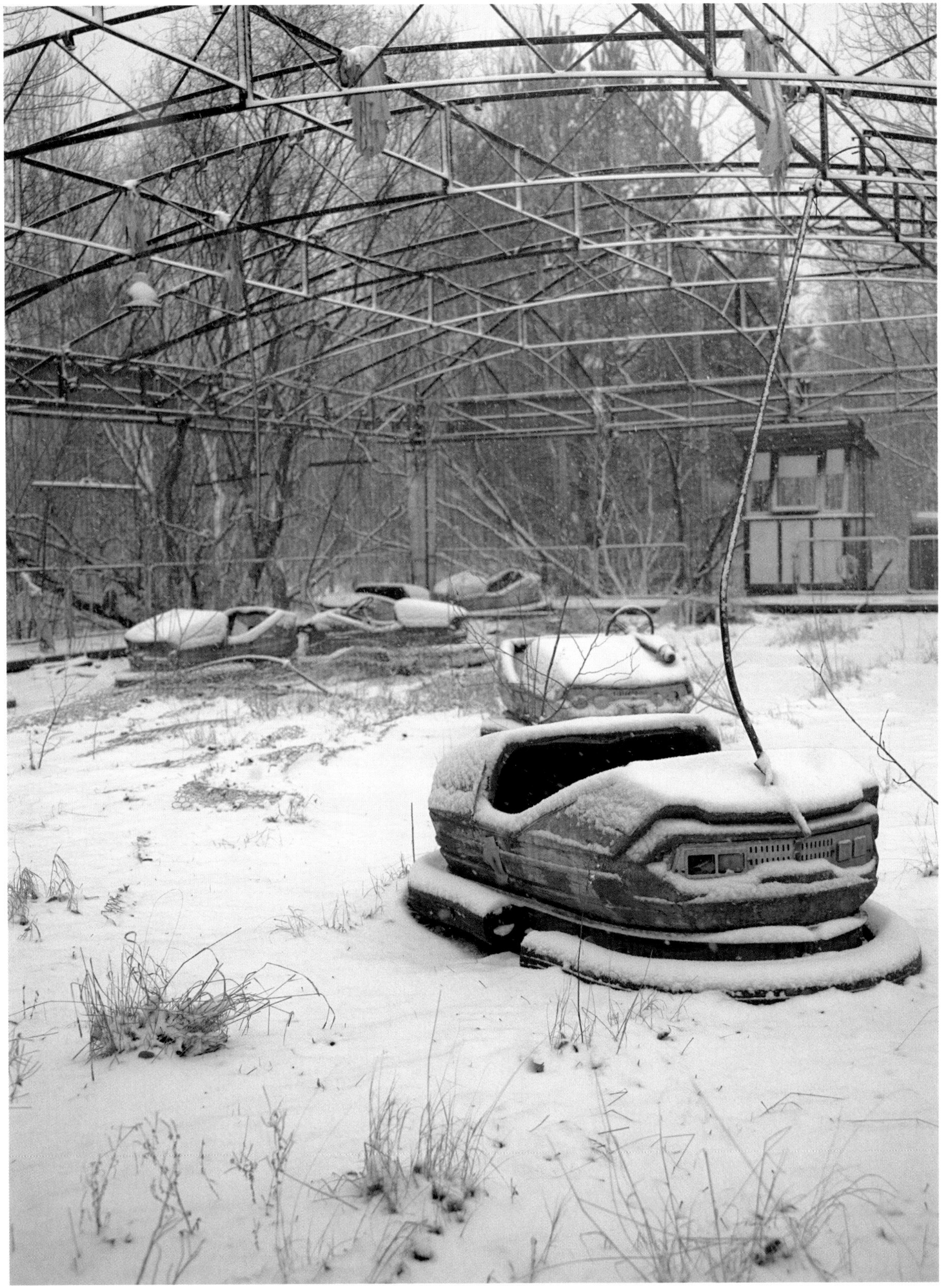

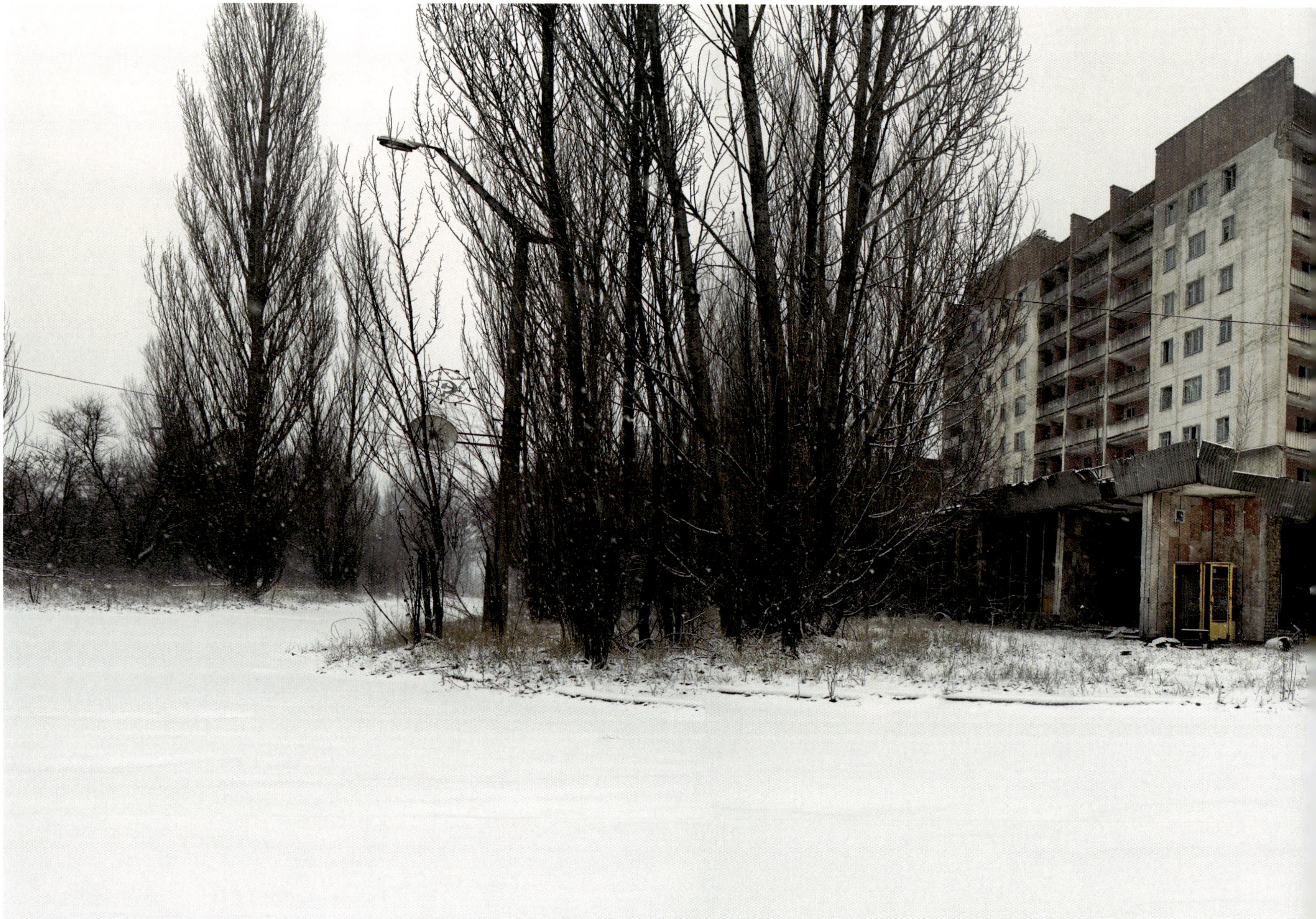

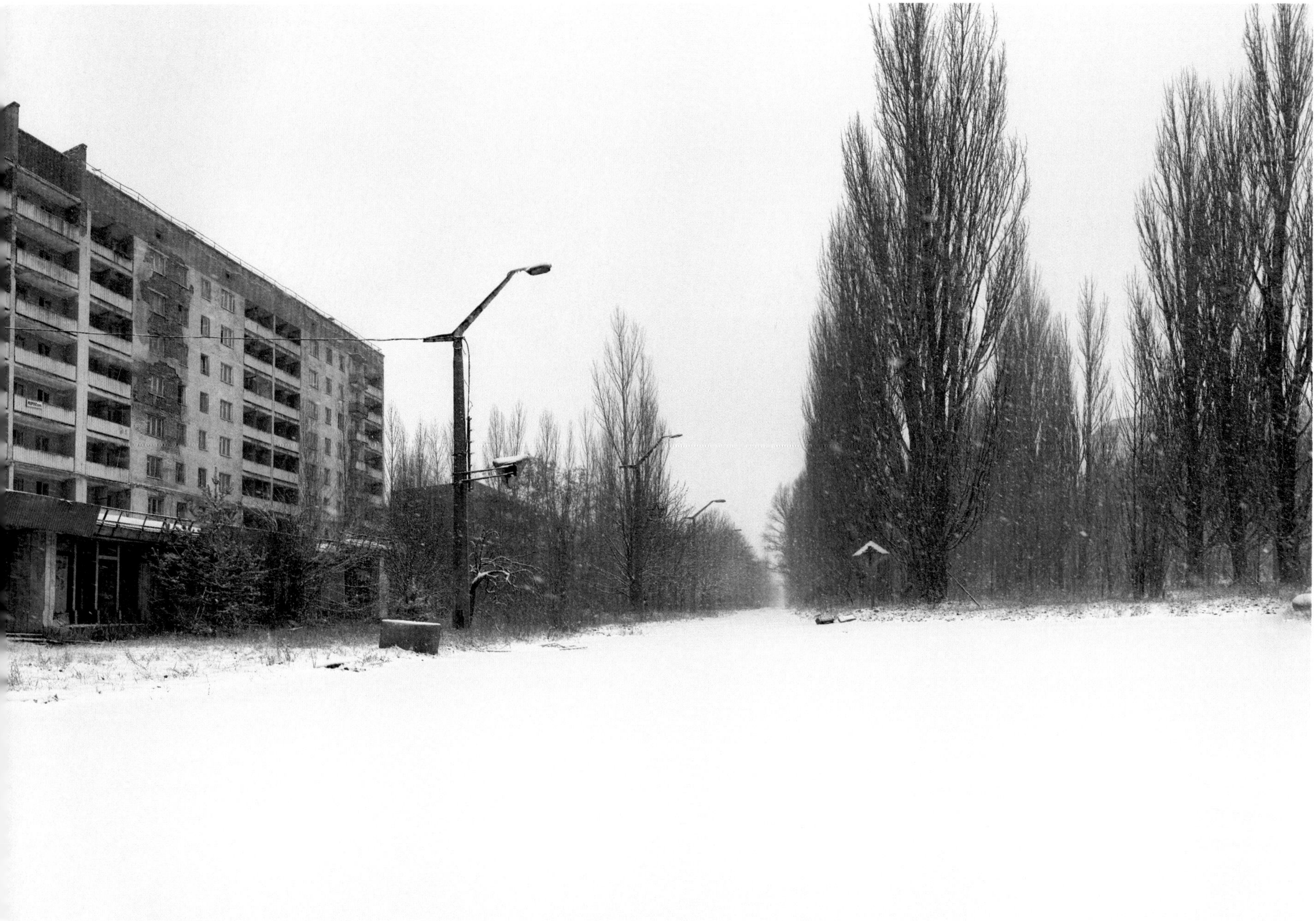

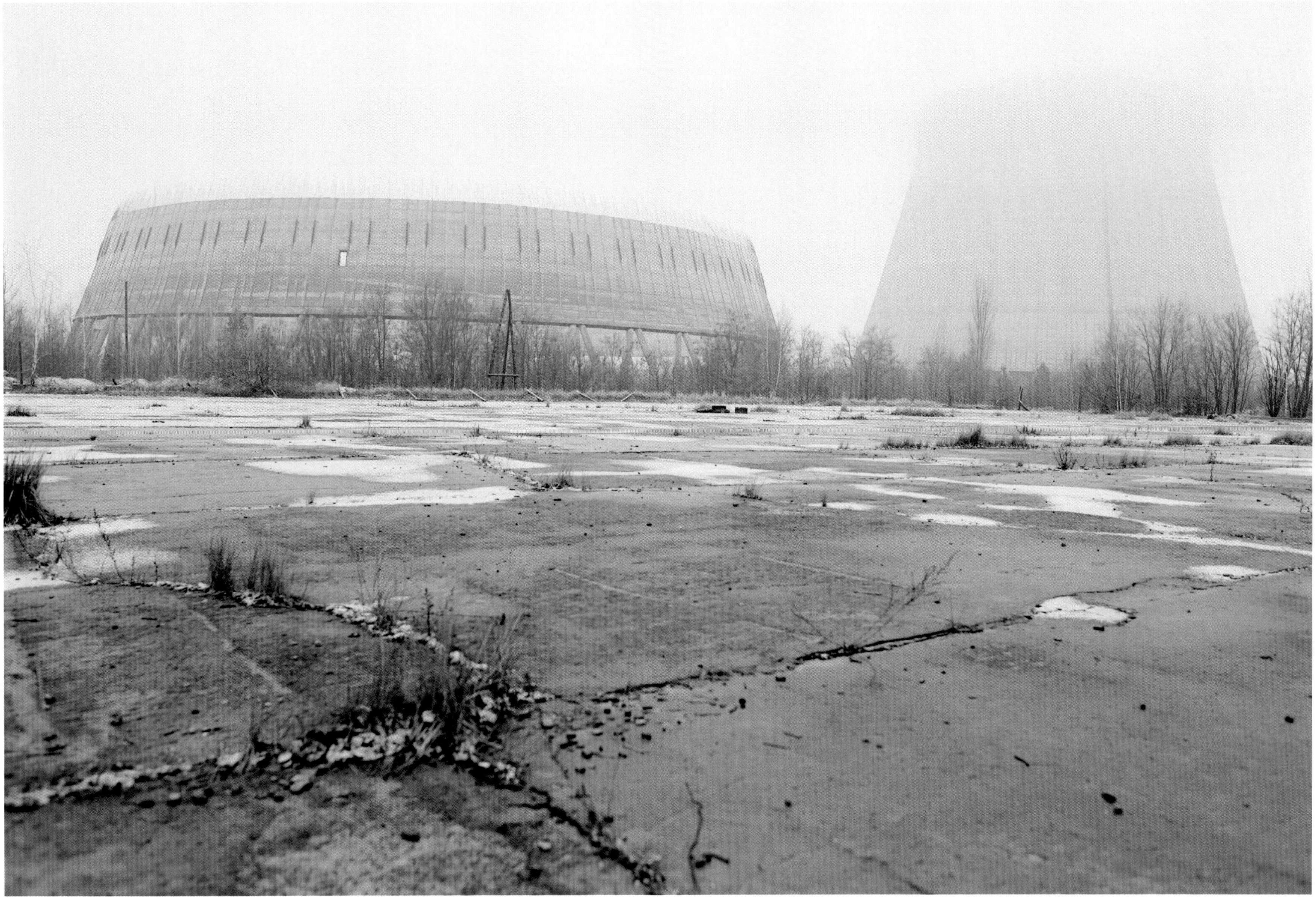

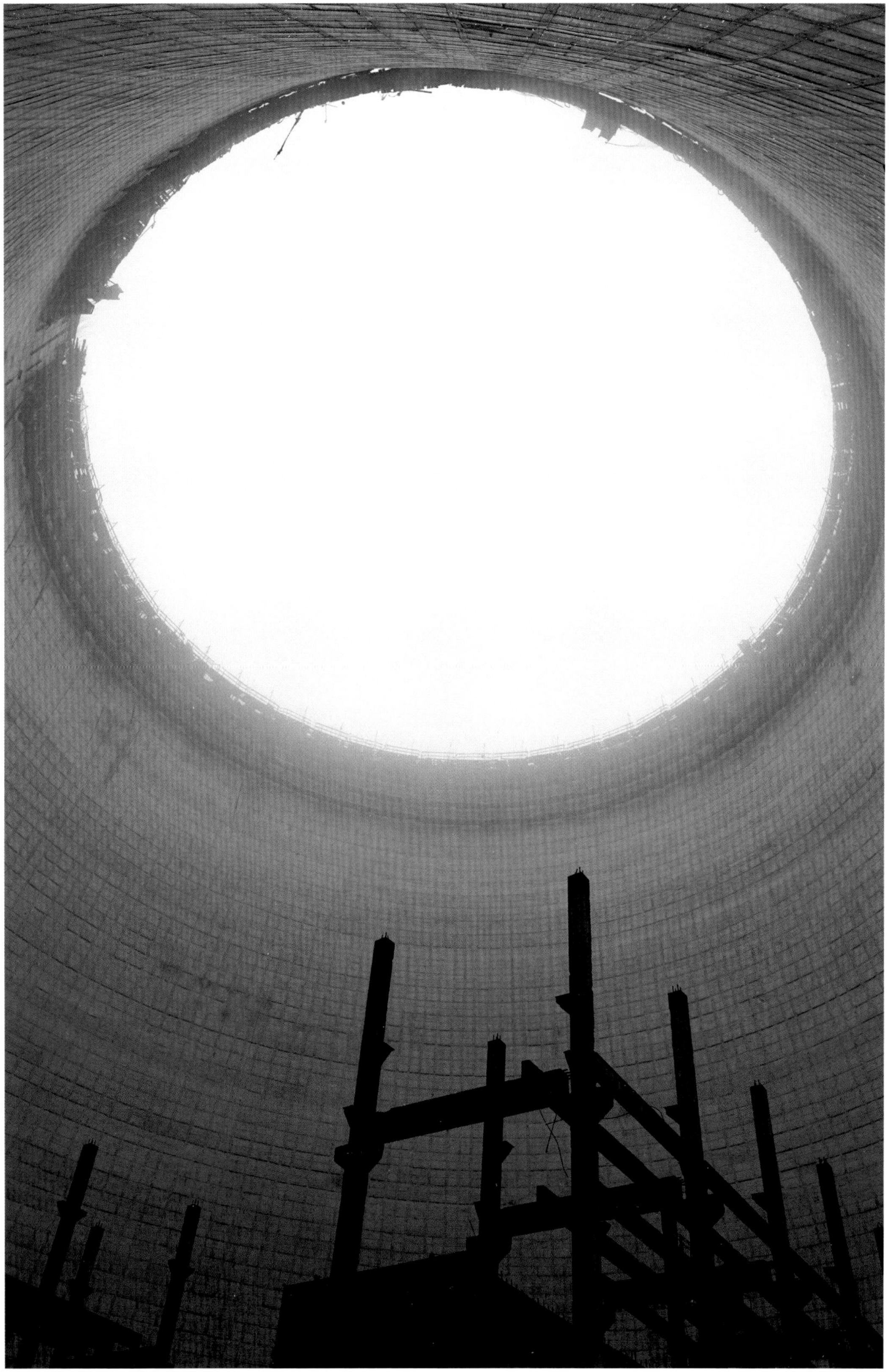

# APPENDIX

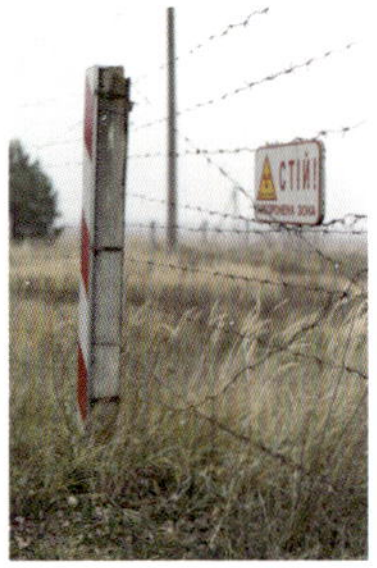

"Untitled", from the series
"Journey into the Future. Stop #1"
C-print, 50 x 75 cm
Page / Seite 17

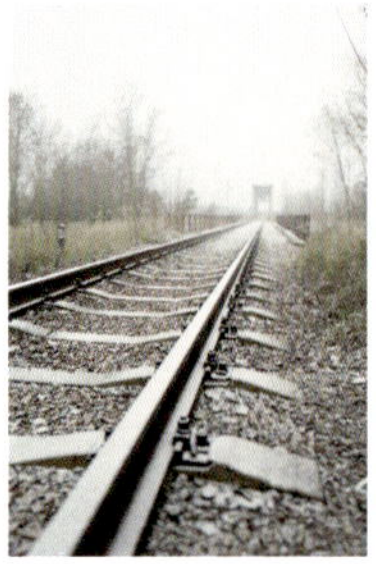

"Untitled", from the series
"Journey into the Future. Stop #1"
C-print, 50 x 75 cm
Page / Seite 19

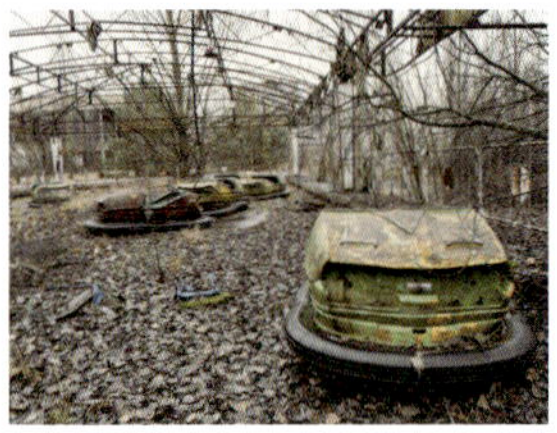

"Untitled", from the series
"Journey into the Future. Stop #1"
C-print, 75 x 56 cm
Page / Seite 21

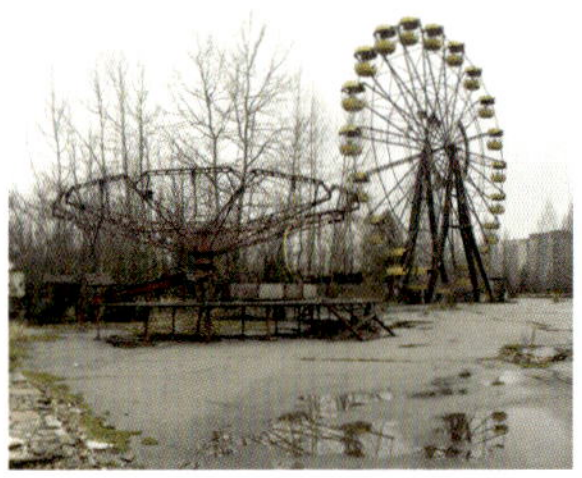

"Untitled", from the series
"Journey into the Future. Stop #1"
C-print, 144 x 120 cm
Page / Seite 23

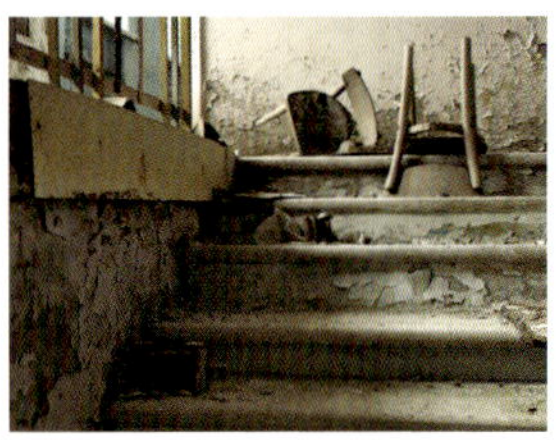

"Untitled", from the series
"Journey into the Future. Stop #1"
C-print, 75 x 56 cm
Page / Seite 25

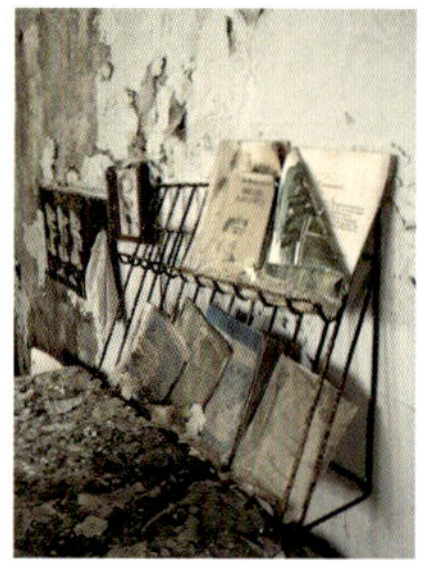

"Untitled", from the series
"Journey into the Future. Stop #1"
C-print, 56 x 75 cm
Page / Seite 27

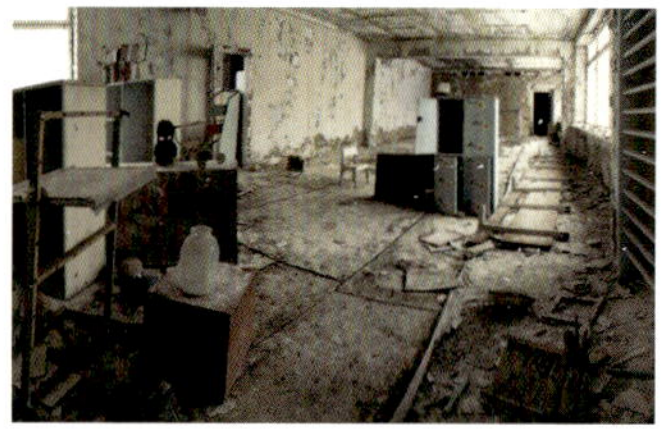

"Untitled", from the series
"Journey into the Future. Stop #1"
C-print, 186 x 120 cm
Page / Seite 29

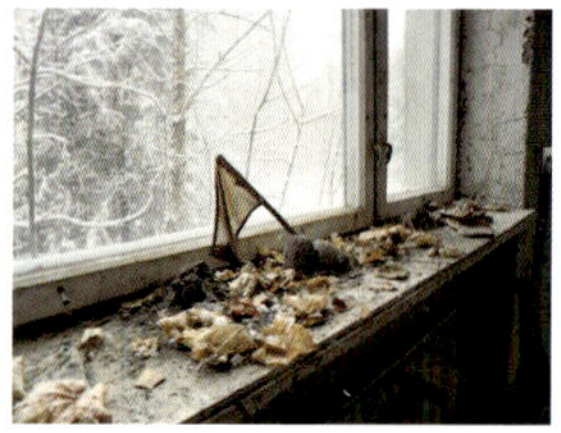

"Untitled", from the series
"Journey into the Future. Stop #1"
C-print, 160 x 120 cm
Page / Seite 31

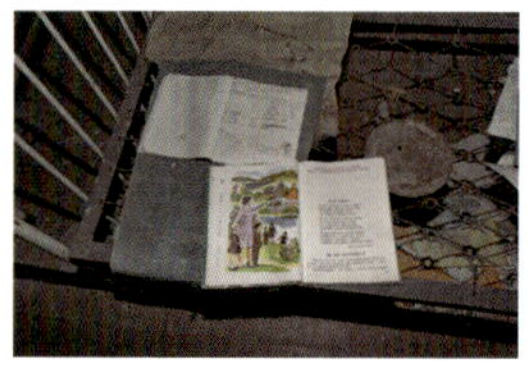

"Untitled", from the series
"Journey into the Future. Stop #1"
C-print, 75 x 50 cm
Page / Seite 33

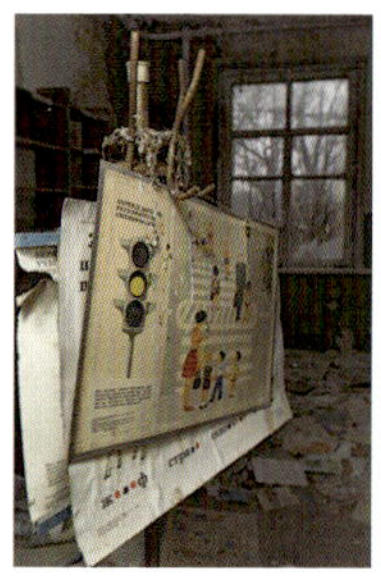

"Untitled", from the series
"Journey into the Future. Stop #1"
C-print, 50 x 75 cm
Page / Seite 35

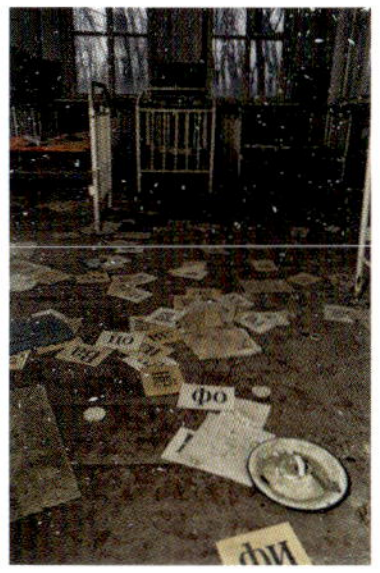

“Untitled”, from the series
“Journey into the Future. Stop #1”
C-print, 50 x 75 cm
Page / Seite 37

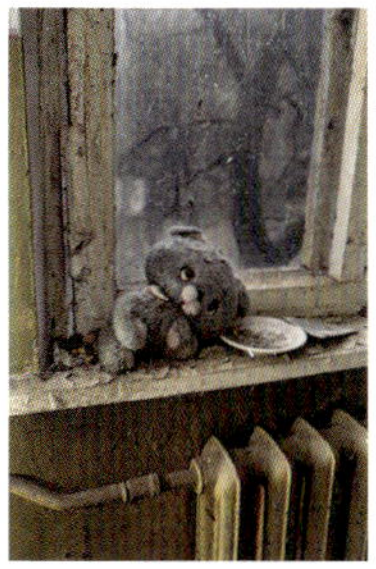

“Untitled”, from the series
“Journey into the Future. Stop #1”
C-print, 120 x 180 cm
Page / Seite 39

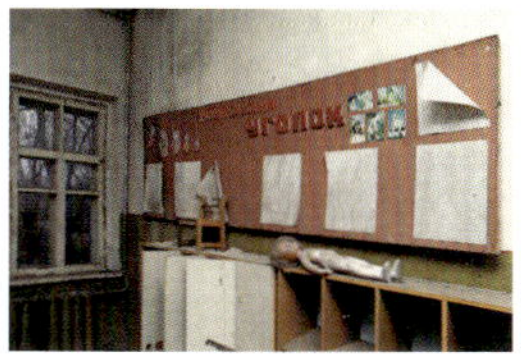

“Untitled”, from the series
“Journey into the Future. Stop #1”
C-print, 177 x 120 cm
Page / Seite 41

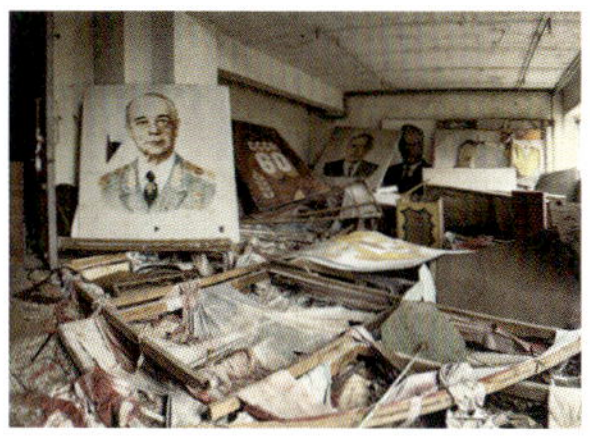

“Untitled”, from the series
“Journey into the Future. Stop #1”
C-print, 168 x 120 cm
Page / Seite 43

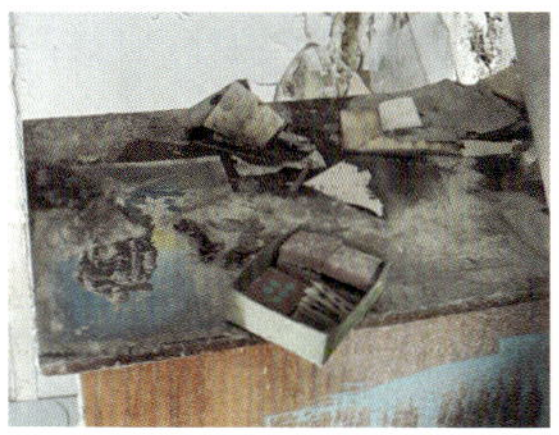

“Untitled”, from the series
“Journey into the Future. Stop #1”
C-print, 75 x 56 cm
Page / Seite 45

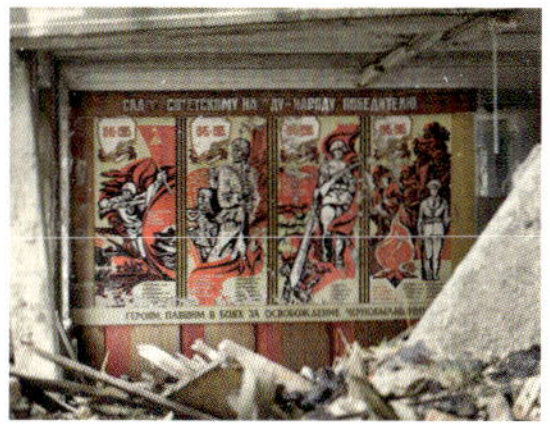

“Untitled”, from the series
“Journey into the Future. Stop #1”
C-print, 75 x 56 cm
Page / Seite 47

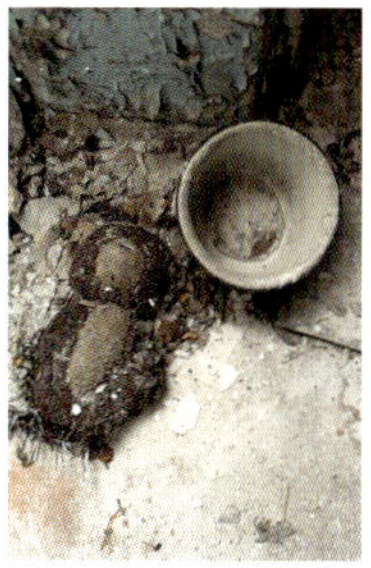

“Untitled”, from the series
“Journey into the Future. Stop #1”
C-print, 50 x 75 cm
Page / Seite 49



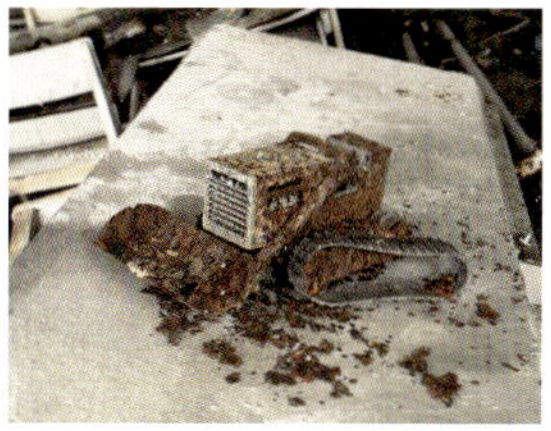

“Untitled”, from the series
“Journey into the Future. Stop #1”
C-print, 75 x 56 cm
Page / Seite 51

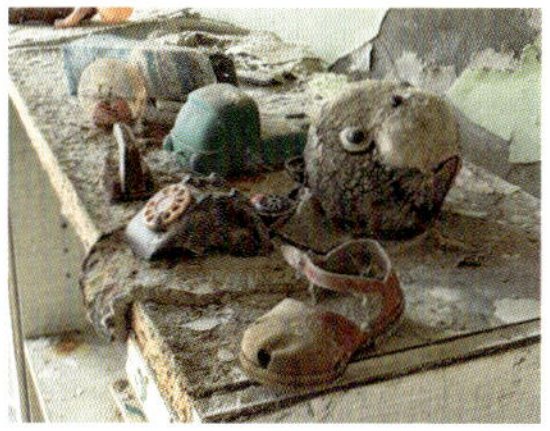

“Untitled”, from the series
“Journey into the Future. Stop #1”
C-print, 75 x 56 cm
Page / Seite 53

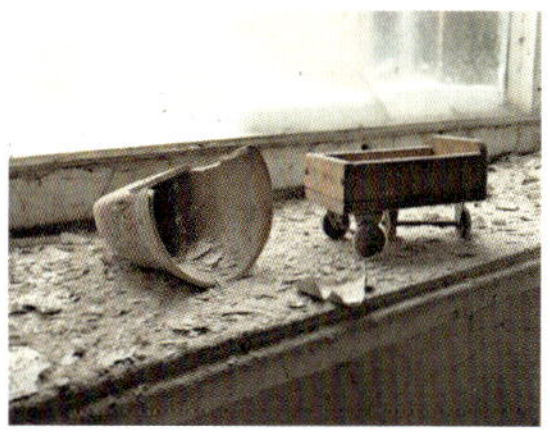

“Untitled”, from the series
“Journey into the Future. Stop #1”
C-print, 75 x 56 cm
Page / Seite 55

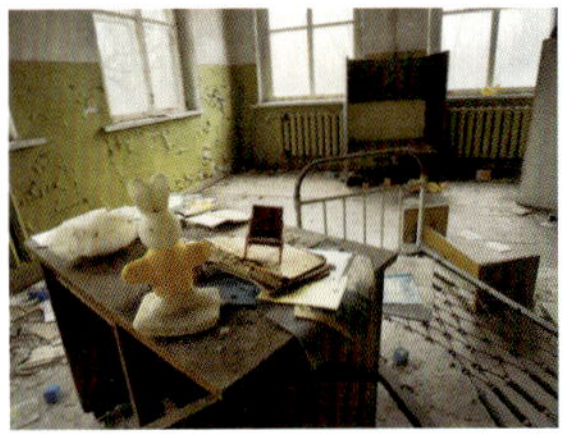

“Untitled”, from the series
“Journey into the Future. Stop #1”
C-print, 75 x 56 cm
Page / Seite 57

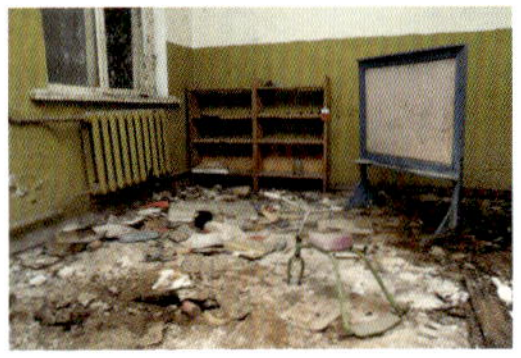

“Untitled”, from the series
“Journey into the Future. Stop #1”
C-print, 75 x 50 cm
Page / Seite 59

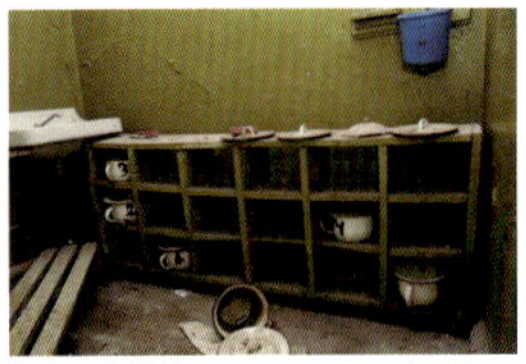

“Untitled”, from the series
“Journey into the Future. Stop #1”
C-print, 180 x 120 cm
Page / Seite 61

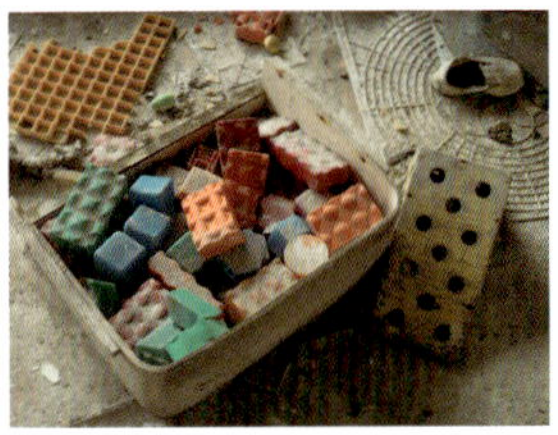

“Untitled”, from the series
“Journey into the Future. Stop #1”
C-print, 75 x 56 cm
Page / Seite 63

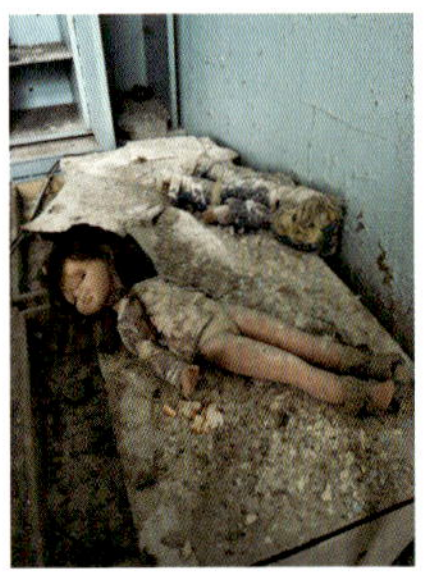

“Untitled”, from the series
“Journey into the Future. Stop #1”
C-print, 75 x 56 cm
Page / Seite 65

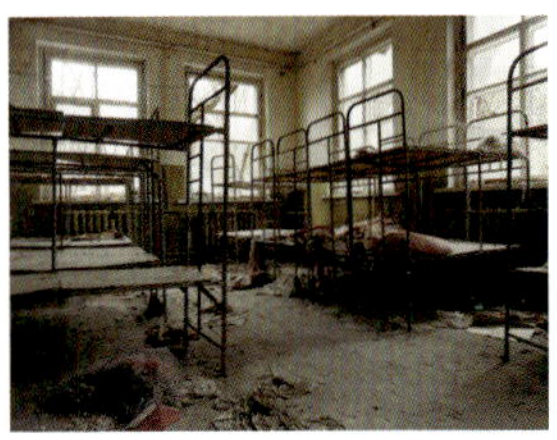

“Untitled”, from the series
“Journey into the Future. Stop #1”
C-print, 75 x 56 cm
Page / Seite 67

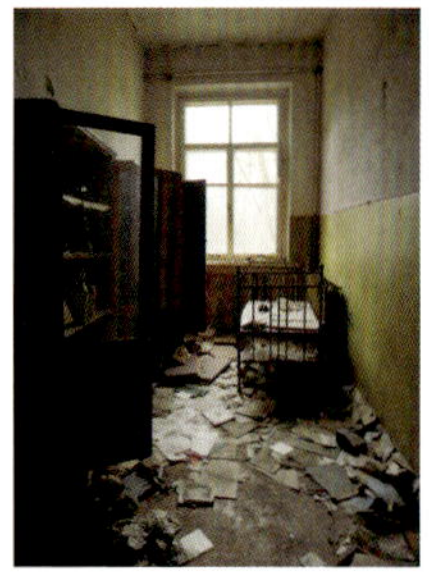

“Untitled”, from the series
“Journey into the Future. Stop #1”
C-print, 120 x 160 cm
Page / Seite 69

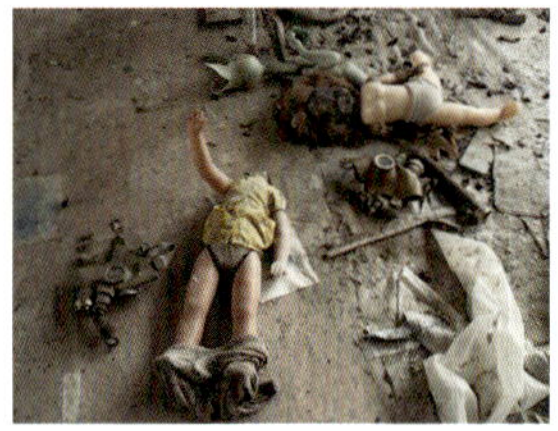

“Untitled”, from the series
“Journey into the Future. Stop #1”
C-print, 75 x 56 cm
Page / Seite 71

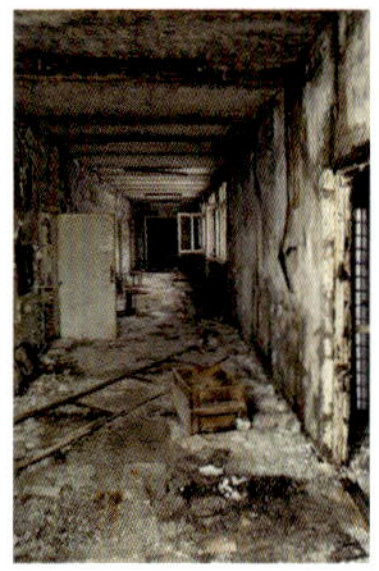

“Untitled”, from the series
“Journey into the Future. Stop #1”
C-print, 50 x 75 cm
Page / Seite 73

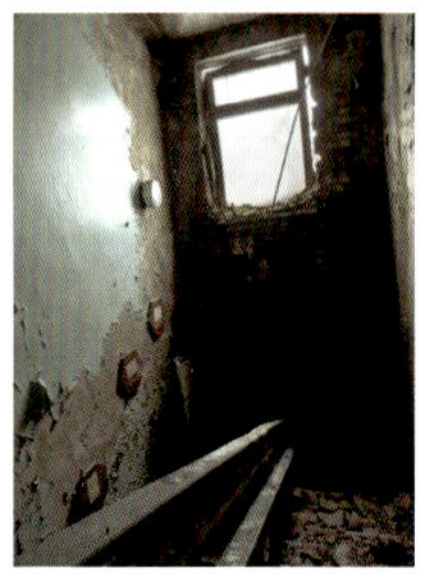

“Untitled”, from the series
“Journey into the Future. Stop #1”
C-print, 56 x 75 cm
Page / Seite 75

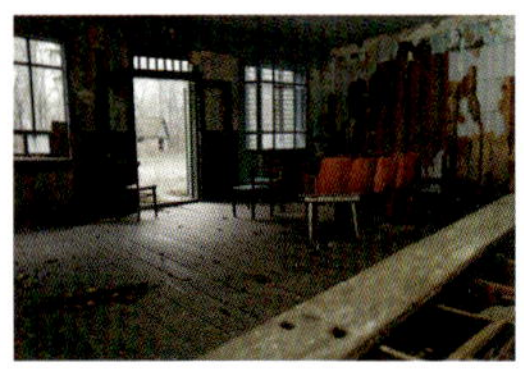

“Untitled”, from the series
“Journey into the Future. Stop #1”
C-print, 180 x 120 cm
Page / Seite 77

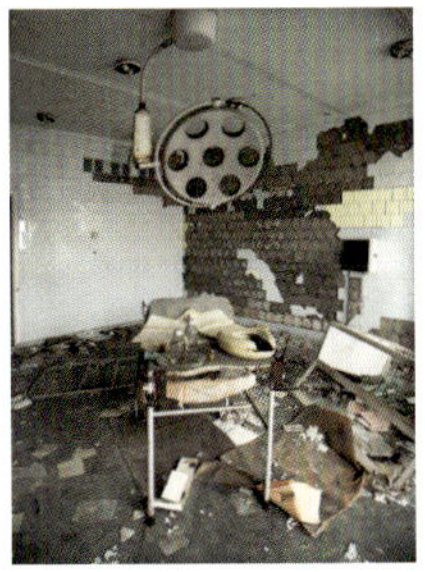

“Untitled”, from the series
“Journey into the Future. Stop #1”
C-print, 56 x 75 cm
Page / Seite 79

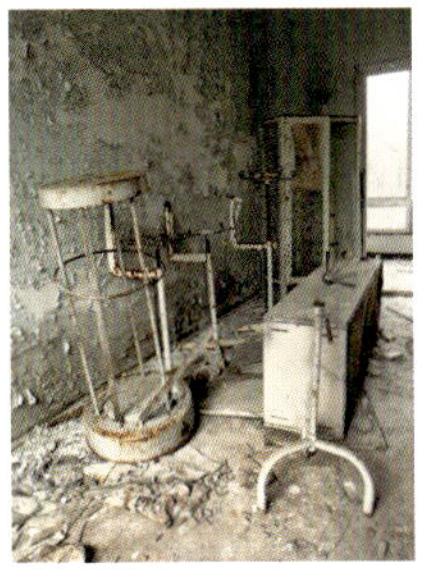

“Untitled”, from the series
“Journey into the Future. Stop #1”
C-print, 56 x 75 cm
Page / Seite 81

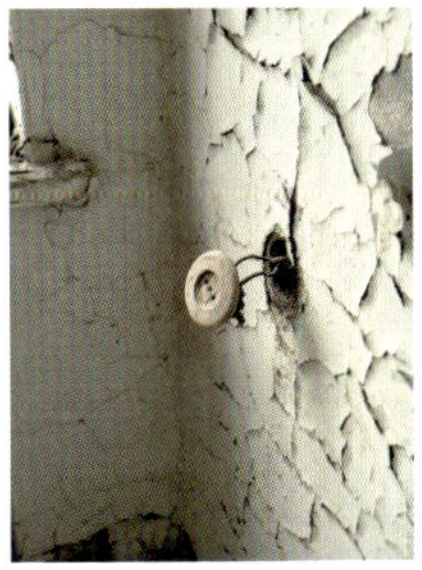

“Untitled”, from the series
“Journey into the Future. Stop #1”
C-print, 56 x 75 cm
Page / Seite 83

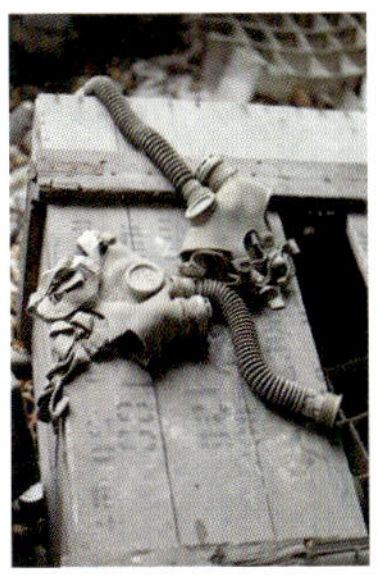

“Untitled”, from the series
“Journey into the Future. Stop #1”
C-print, 50 x 75 cm
Page / Seite 85

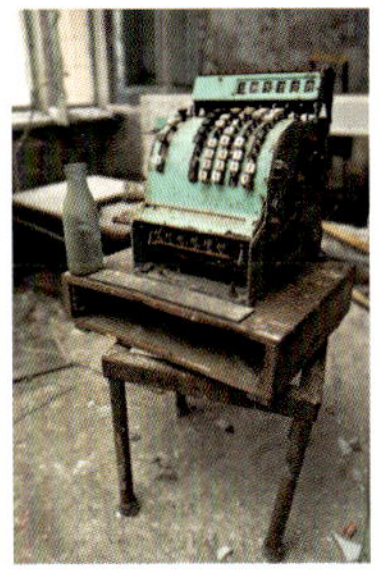

“Untitled”, from the series
“Journey into the Future. Stop #1”
C-print, 50 x 75 cm
Page / Seite 87

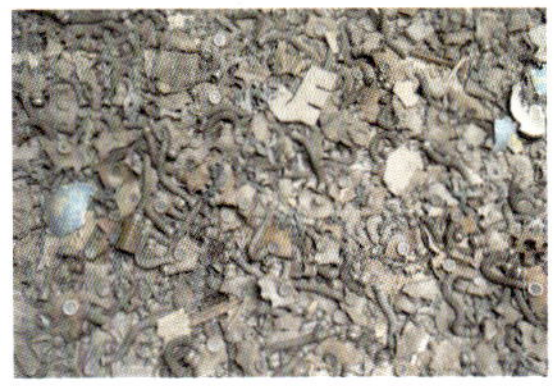

“Untitled”, from the series
“Journey into the Future. Stop #1”
C-print, 180 x 120 cm
Page / Seite 89



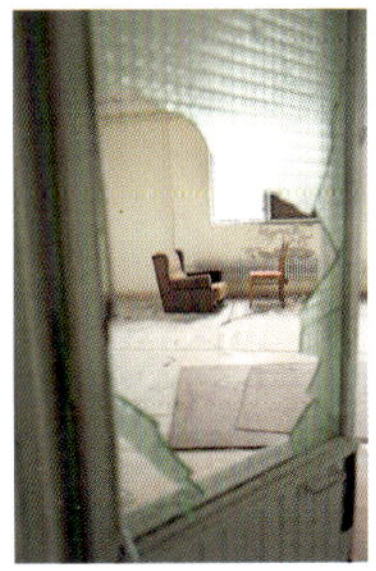

“Untitled”, from the series
“Journey into the Future. Stop #1”
C-print, 50 x 75 cm
Page / Seite 91

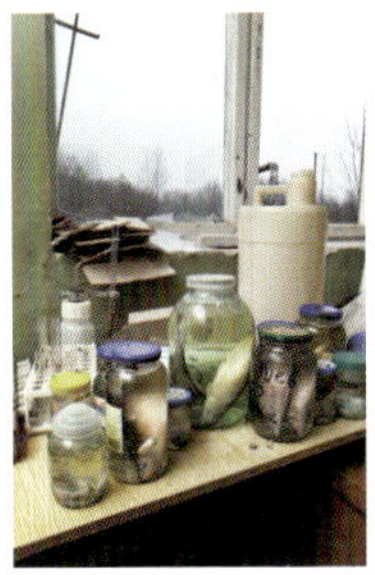

“Untitled”, from the series
“Journey into the Future. Stop #1”
C-print, 50 x 75 cm
Page / Seite 93

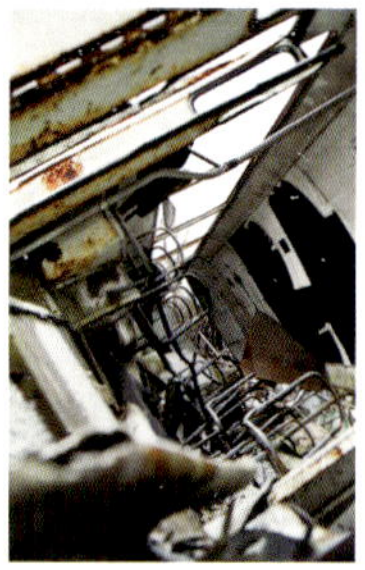

"Untitled", from the series
"Journey into the Future. Stop #1"
C-print, 50 x 75 cm
Page / Seite 95

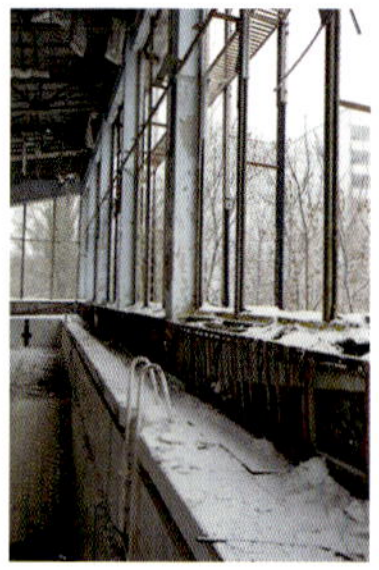

"Untitled", from the series
"Journey into the Future. Stop #1"
C-print, 50 x 75 cm
Page / Seite 97

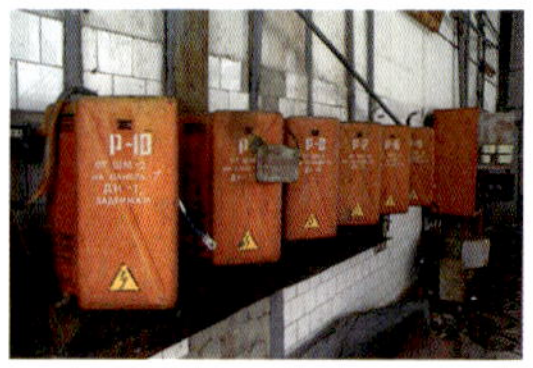

"Untitled", from the series
"Journey into the Future. Stop #1"
C-print, 75 x 50 cm
Page / Seite 99

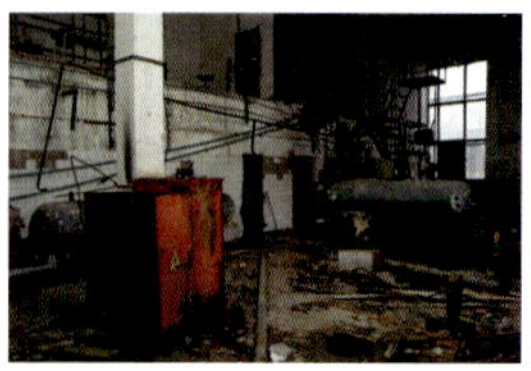

"Untitled", from the series
"Journey into the Future. Stop #1"
C-print, 180 x 120 cm
Page / Seite 101

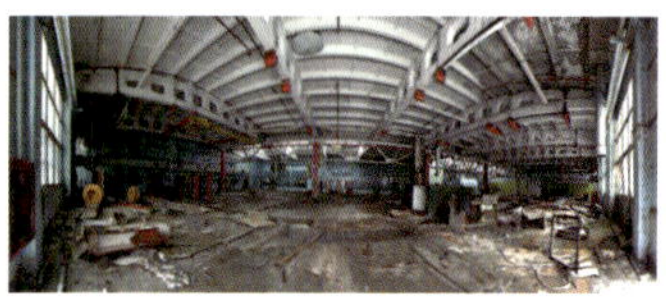

"Untitled", from the series
"Journey into the Future. Stop #1"
C-print, 180 x 74 cm
Page / Seite 103

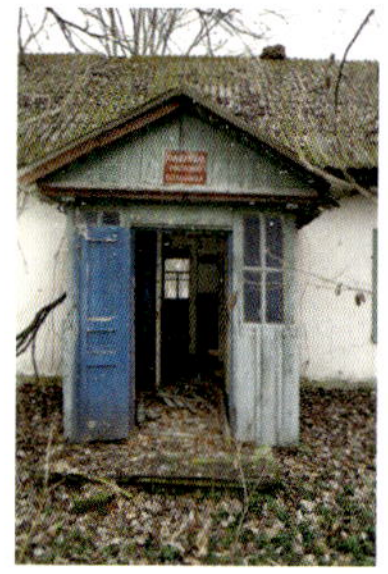

"Untitled", from the series
"Journey into the Future. Stop #1"
C-print, 50 x 75 cm
Page / Seite 105

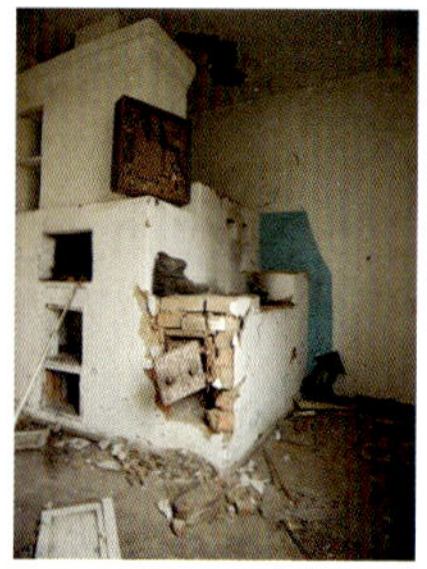

"Untitled", from the series
"Journey into the Future. Stop #1"
C-print, 56 x 75 cm
Page / Seite 107

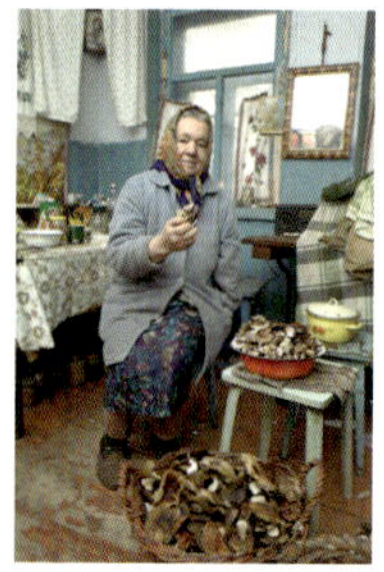

"Untitled", from the series
"Journey into the Future. Stop #1"
C-print, 50 x 75 cm
Page / Seite 109

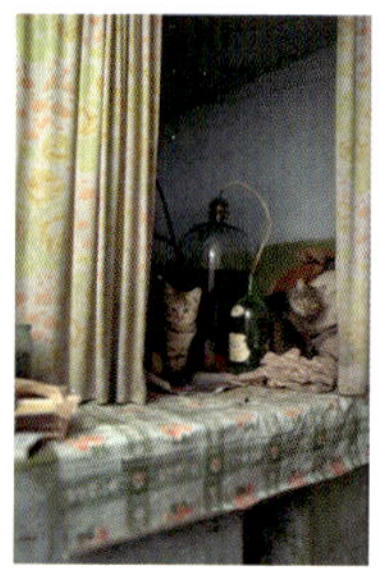

"Untitled", from the series
"Journey into the Future. Stop #1"
C-print, 50 x 75 cm
Page / Seite 111

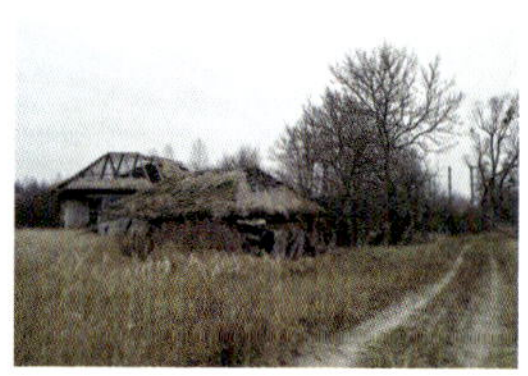

"Untitled", from the series
"Journey into the Future. Stop #1"
C-print, 160 x 120 cm
Page / Seite 113

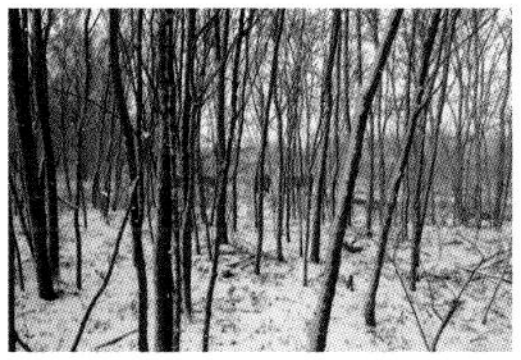

"Untitled", from the series
"Journey into the Future. Stop #1"
C-print, 75 x 50 cm
Page / Seite 115

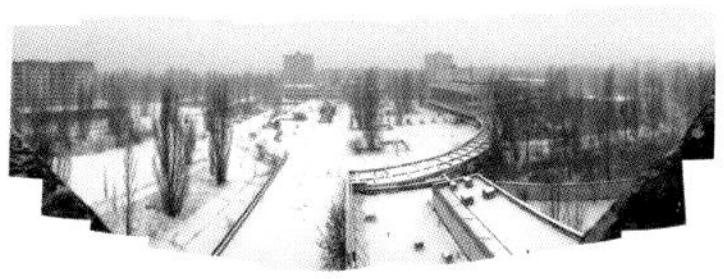

"Untitled", from the series
"Journey into the Future. Stop #1"
C-print, 75 x 28 cm
Page / Seite 117

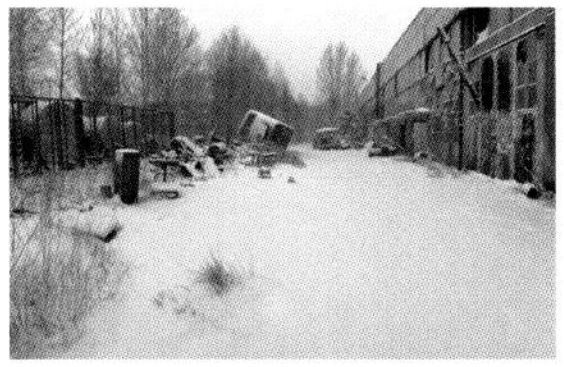

"Untitled", from the series
"Journey into the Future. Stop #1"
C-print, 75 x 48 cm
Page / Seite 119

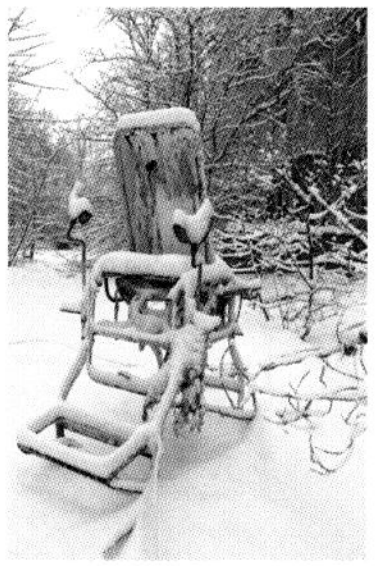

"Untitled", from the series
"Journey into the Future. Stop #1"
C-print, 50 x 75 cm
Page / Seite 121

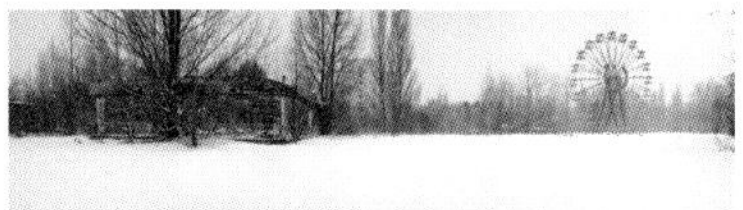

"Untitled", from the series
"Journey into the Future. Stop #1"
C-print, 180 x 50 cm
Page / Seite 123

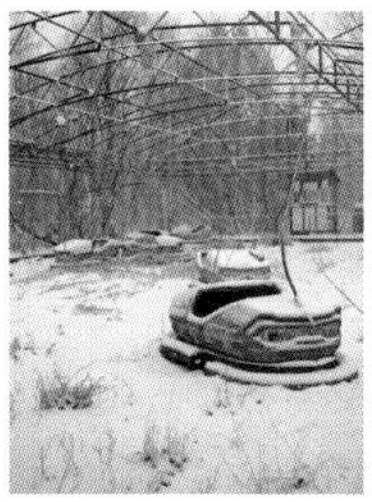

"Untitled", from the series
"Journey into the Future. Stop #1"
C-print, 56 x 75 cm
Page / Seite 125

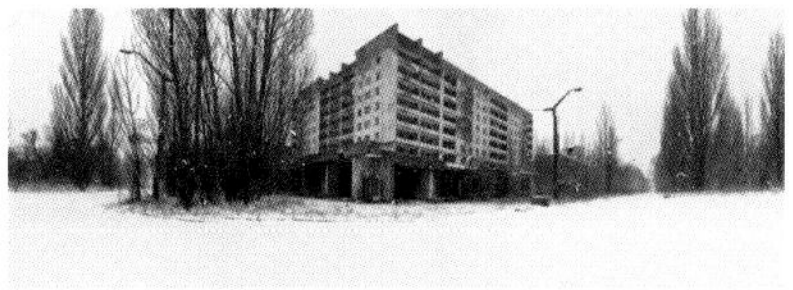

"Untitled", from the series
"Journey into the Future. Stop #1"
C-print, 180 x 64 cm
Page / Seite 127

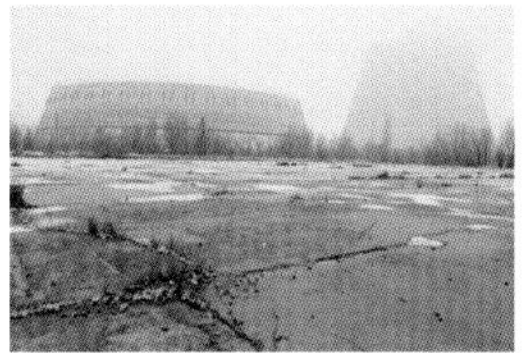

"Untitled", from the series
"Journey into the Future. Stop #1"
C-print, 75 x 50 cm
Page / Seite 129

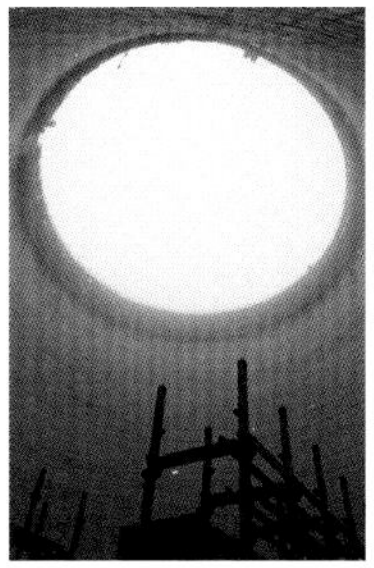

"Untitled", from the series
"Journey into the Future. Stop #1"
C-print, 50 x 75 cm
Page / Seite 131

# BIOGRAPHY



## SERGEY SHESTAKOV

Born in 1968.
Lives and works in Moscow.

### STUDIES

1985-1991
Moscow Physico-Technical Institute

### SOLO EXHIBITIONS

2011
Bhutan. Project “Happiness”. The State Museum of Oriental Art, Moscow
Territory_Pole. Russian State Arctic and Antarctic Museum, St.Petersburg
En marche vers le future. Galerie L’Aleatoire, Paris
Journey into the future. Stop #1. Multimedia Art Museum, Moscow
Extra poles. Mosfilm Gallery, Moscow

### SELECTED GROUP EXHIBITIONS

2011
Antarctic Biennale International Art Exhibition. “Ultima Thule”.
Ca’Foscari University, Venice

2010
The Best of Russia’10. Center of Modern Art “Vinzavod”, Moscow

# German Translation
# Deutsche Übersetzung

## VORWORT

von Sergey Shestakov

Tschernobyl ... Dieses Wort ist schon längst zu einem Appellativum geworden - zum Symbol einer schrecklichen menschlichen Tragödie. Unser Gedächtnis funktioniert so, dass im Verlauf der Zeit die Details verblassen, die Empfindungen verstummen und alles, was geschehen ist die Konturen verliert, verschwommen wird, so als ob jemand die Schärfe des Objektivs der Geschichte verstellt. So ist es auch den Ereignissen in Tschernobyl ergangen, jenen Ereignissen, die einst die Welt erschüttert haben. Während des Vierteljahrhunderts, welches seit der Tschernobylkatastrophe vergangen ist, schien uns die Katastrophe, die sich vom v. auf den 26. April 1986 ereignet hat, als etwas absolut Schreckliches, das nur unter bestimmten schlechten, außerordentlichen Umständen passieren konnte. Es schien uns, dass so etwas Schreckliches niemals wieder passieren könne. Mit dem 11. März 2011 hat sich jedoch alles verändert, als die zehn Meter große Welle, verursacht durch das starke Erdbeben, dem Atomkraftwerk Fukushima 1 enorme Schäden zufügte.

Als ich im Oktober die Reise zu der 30 km langen, unbewohnbaren Zone bei Tschernobyl antrat, ahnte ich nicht, dass die Fotografien, die ich dort machen würde, zu einer Ausstellung gelangen und gar in einem Buch veröffentlicht werden würden. Die Aufnahmen sollten der Beginn eines großen Fotoprojekts sein, das voraussichtlich 2012 beendet werden soll; das Nachdenken über das Schicksal der Menschen ist dabei zentral: Prypjat wurde eher zufällig gewählt. Das,was ich dort sah, hat mich zutiefst erschüttert. Wir (ich und mein Fremdenführer) liefen durch eine absolut leere Stadt. Das Gefühl, dass gleich die Geräusche von vorbeifahrenden Autos oder das Lachen der spielenden Kinder zu hören wären oder spazieren gehende junge Mütter mit Kinderwagen auftauchen würden, ließ mich nicht los. Aber die absolute Stille wurde nicht mal durch Vogelgezwitscher gestört. Je mehr wir uns fortbewegten, um so stärker wurde das Gefühl, dass alles, was wir sahen irreal war. Dieses rrealitaÅNtsgefühl wurde nicht vom Gedanken begleitet, dass so etwas gar nicht realistisch sein kann, sondern, dass dies NICHT SEIN DARF. Städte und Dörfer dürfen nicht leer stehen, die Umwelt darf nicht zerstört werden, die Menschen dürfen nicht sich selbst vernichten.

Die Idee mit dem Titel der Ausstellung kam mir in den Sinn, als ich in einem der Prypiatskindergärten das halbverfallene Buch von Zoya Voskresenskaja sah: „Die Fahrt in die Zukunft". Ich habe nur den Untertitel „Station Nummer 1" hinzugefügt, da Tschernobyl keine Endstation ist. Heute sind die Menschen auf der Station Nummer 2 ausgestiegen und sollen sich jetzt entscheiden, auf welcher Linie der „Lebensbahn" sie ihre Reise fortsetzen.

Alles liegt in unseren Händen ...

## KUNST DER ZUKUNFT. TCHERNOBYL FOTOGRAFIERT VON SERGEY SHESTAKOV

von Wakana Kono

„Zukunftsbesessenheit ist eine der Verführungen und der Flüche des 20. Jahrhunderts – ein Erbe des Optimismus und Fortschritsgedankens des 19. Jahrhunderts. Es ist aber kein Grund dafür, das 21. Jahrhundert mit der vom 20. Jahrhundert vererbten Skepsis zu vergiften. Wo die Gefahr herkommt, daher kommt auch die Rettung.“
Michail Epstein. Aus dem Zyklus „Neues Modell der Zukunft“
Aus der Serie 'To the New Model of the Future'. Artikel 1[1]

Als Michail Eptein die heutige Zeit als Postmoderne bezeichnet hat, meinte er den Zerfall von der UdSSR, den Zerfall des Kommunismus, der seiner Meinung nach eine Form der „Futuromanie“ bzw. „Futurokratie“[2] war. Tatsächlich ist das Ende der utopischen und antiutopischen Ideen darunter auch zugleich das Ende des Glaubens an den „friedlichen Atom“ eine der grundlegenden Tendenzen der modernen Zeit.

Zeitgenössische Künstler stellen nach der Suche der Zukunftsmodelle der Postmoderne unterschiedliche Weltgestalten vor. Konstantin Batynkov z.B. in seinem graphischen Zyklus „Auf den Wellen Laufende“ gestaltet die Katastrophen der Postmoderne. Im Vordergrund der Kriegs- und Umweltkatastrophen seiner Zeichnungen ist eine auf den Wellen laufende Mädchensilouette zu sehen. Leonid Tishkov zeigt mit seinen Figuren von „Dabloiden“ und „Tauchern“ und dem ideellen Raum „Ladomir“ die Möglichkeit einer utopischen Welt. Stanislav Shuripa, der Künstler der jungen Generation, die in der Zeit der Computerisierung aufgewachsen sind, stellt den virtuellen Raum mit der Technik der Ölmalerei dar, auf der Suche nach einer neuen Sprache zur Beschreibung der neuen Welt.

Aleksander Ponamorev und Sergey Shestakov haben vor kurzem eine Fotoausstellung unter dem Titel „Zusätzliche Pole“ eröffnet, auf der die glänzenden Erdpole gezeigt wurden. Die Ausstellung soll eine Ankündigung auf die bevorstehende unglaubliche, nie dagewesene Internationale Biennale der zeitgenössischen Kunst sein. Nach der Meinung von Shestakov wird diese in der Antarktika organisierte Biennale die internationalen zeitgenössischen Künstler vereinen, wird ihnen eine gemeinsame kreative Arbeit ermöglichen. Es ist sehr symbolisch, da die Antarktika eigentlich der einzige Ort auf der Welt ist, wo es keine Grenzen gibt. So wie die moderne Kunst keine Grenzen hat und alle Barrieren und Stereotypen überwindet und dabei neue Energien und Inhalte schafft. Dieses Projekt soll motivierend auf die zukünftige Kunst einer Welt „ohne Grenzen“ sein. „Die Fahrt in die Zukunft. Station Nr.1“ von Shestakov stellt auf andere Weise die Landschaften der Postmoderne dar. In diesem Zyklus stellt Shestakov die Tschernobylfotografien vor, die er 2010 in der 30 km langen unbewohnbaren Zone gemacht hat. Die Idee der Namensgebung des Zyklus kam ihm in den Sinn, als er in einem der verlassenen Kindergärten Pripyats das Buch von Zoya Voskresenskaja „Die Fahrt in die Zukunft“ sah. Dieser Titel trifft hervorragend die Botschaft des Zyklus: Die Fotografien der Landschaft der ehemaligen Stadt spiegeln Vergangenheit, Gegenwart und Zukunft wider. Auf den Tschernobylfotografien von Shestakov kann ein aufmerksamer Betrachter die Situation des Kraftwerks Fokushima 1 und ein mögliches Menschheitsschicksal in der Zukunft wiedererkennen. Indem Shestakov die Werke über die Vergangenheit schafft, spricht er über die Krisensituation der heutigen Zeit und das Schicksal der Menschen. Dieses ruft Assoziationen der Ähnlichkeit mit der totalen Installation von Ilya Kabakov hervor, in der er den sowjetischen Raum wiederherstellte – „10 Personen“ und „In der kommunalen Küche“ von Kabakov stellen nicht nur die sowjetische Wirklichkeit dar, sondern spiegeln den globalen Zustand der heutigen Zeit wider.

Die Aufnahmen von Häuser und Straßen von Pripyat, gefüllt mit verlassenen Gegenständen, vergilbten Heften, staubigen Spielzeugen ähneln in einer gewissen Weise den Installationen von Kabakov, vor allem an die „Toilette“ und „Schule Nr. 6“. In der Reihe der vielen Tschernobylfotografien verschiedener Autoren stechen die Fotografien von Shestakov durch ihre Emotionalität hervor – und genau durch diese Eigenschaft sind sie den Installationen von Kabakov so nah. Im Jahre 1990 im Dialog mit Boris Groys sagte Kabakov “[...] ich empfinde die ganze Welt und mich selbst, gefüllt mit verschiedenen Klängen: Knattern, Plappern, Schnalzen, Türenknallen, Autogetöse. Dies alles ist der Klang des Seins. Darunter sind die Stimmen der Menschen, welche mich umgeben, welche in mir erklingen. Die ganze Zeit erfüllt mich der Wunsch, auf diese Stimmen zu antworten, die eigene Existenz zu manifestieren.“[3] „ [...] Wenn ich etwas sehe, fängt es an in mir zu erklingen. Es vollzieht sich eine Wiedergeburt des Klanges aus dem toten Gegenstand.“[4] Genau dieses können wir in den Fotografien von Shestakov nachempfinden: Wir hören die Stimmen der ehemaligen Bewohner, das Vorlesen der Lehrer, Schreie von Sportler. Dort erklingen alltägliche Geräusche: Die Mutter beim Zubereiten des Abendessens, der Nachbar, der die Tür öffnet, der Frühlingswind, welcher weht. Es ist aber absolute Stille in der Stadt – der Blick von Shestakov, seine schöpferische Überarbeitung erfüllen die Fotografien mit den Lebensklängen und den Bruchteilen der Erinnerung an die Verschwundenen. Wir als Betrachter können die Empfindung des Künstlers nachvollziehen.

Man kann noch eine Parallele ziehen zu den japanischen Künstler Kendji Yanobe, der sich seit Anfang der 90er mit Media Art, Fotografien und Perfomances beschäftigt. Yanobe ist 1965 in Osaka geboren, wo 1970 die Weltausstellung stadtfand. Er spielte oft als Kind in den Ruinen dieses Gebäudes, das später für ihn zur Methapher der Apokalypse und des Menschheitsschicksals wurde. Yanobe fing an Werke zum Thema „Wie der Mensch das Ende der Welt überleben kann“ zu schaffen. 1994 war er in Pripyat und hat Häuser und Straßen in der leeren Stadt fotografiert. Diese Reise nach Tschernobyl hat einen erschütternden Eindruck auf ihn hinterlassen: Dort hat er das Ende der Welt in Realität gesehen, so dass er sein Hauptthema vom „Überleben“ in „Neubeginn“ änderte.

Die Gestalt der Sonne auf der Wand eines Kindergartens wurde zum Leitfaden seiner schöpferischen Tätigkeit, die dem Neubeginn und der besseren Zukunft der Kinder gewidmet wurde. Es ist kein Zufall, dass beide – Yanobe und Shestakov die Symbole ihrer Tschernobylzyklen in einem verlassenen Kindergarten gefunden haben: Für Yanobe war es die Sonne, für Shestakov das Buch „Die Fahrt in die Zukunft“. Zu diesem Titel fügte Shestakov „Station Nr.1“ hinzu „da Tschernobyl keine Endstation ist“. In diesem Jahr ist Fukushima zur tragischen Station Nr.2 geworden. Aber die Fotografien von Shestakov, die mit Emotionen, Atem und Stimmen – seiner eigenen sowie jener der Bewohner der Stadt, welche weggegangen sind – geben uns doch die Hoffnung auf die Fahrt in eine bessere Zukunft.

1 http//veer.info/01/v1.bnd.html
2 ebenda
3 Ilya Kabakov, Boris Groys. Dialoge. Wologda: Bibliothek des moskauer Konzeptionalismus. Hermann Titow, 2010, S. 18
4 ebenda S. 19

## EINLEITUNG

von Olga Sviblova

Die meisten Projekte, mit denen ein Kurator oder Museumsdirektor arbeitet, sind das Ergebnis von langen Reflexionsprozessen. Die Grundlage jedes Projekts ist jedoch ein tiefes Gefühl. Es ist nicht lange her, als ich die Fotografien von Sergey Shestakov zum ersten Mal gesehen habe. Ich erinnere mich sehr gut an den emotionalen und existenziellen Schockzustand und an die seltsame stille Faszination, die ich damals erlebt habe.
Mir schien es, ich wüsste fast alles über Tschernobyl, beziehungsweise ziemlich viel. Tschernobyl war eine Tragödie für viele meiner Freunde und wurde beinah zur Tragödie für die Mitglieder meiner Familie. Einige Tage vor der Katastrophe bereitete eine Gruppe von russischen und ukrainischen Dichtern einen Poesieabend in dem Kernkraftwerk von Tschernobyl vor. Als Witz schrieb man aus Kiev nach Moskau „alles wird stattfinden, wenn nichts explodiert". Die radioaktive Strahlung hat einen Teil dieser Gruppe getroffen. Einige aus der Gruppe haben später kranke Kinder bekommen, einige haben die Chance Kinder zu haben überhaupt verloren.

Ein Vierteljahrhundert habe ich alles über Tschernobyl gelesen, was es zu lesen gab. Ich habe Filme, Foto- und Videoreportagen gesehen. Es waren die Zeugen, die das Vergessen der Tragödie unmöglich machten. Die Arbeiten von Sergey Shestakov sind keine Erinnerungen an die Vergangenheit; sie bringen vielmehr den Betrachter zum Fühlen und zum Nachdenken – vor allem über die Zukunft.

Jede Aufnahme von Sergey Shestakov ist eine durchdringende, aus mehreren Details bestehende Metapher. Das ganze Projekt scheint eine künstlerische Zusammenfassung zu sein, die besonders heute, nach der Katastrophe in Japan in einem anderen Licht scheint. In dieser neuen Situation bekommt auch der Titel „Die Fahrt in die Zukunft" einen andren Sinn.
Das Projekt von Sergej Shestakov spricht nicht nur darüber, wie schwierig es ist, die Gefahr von bestehenden Kernkraftwerken bzw. von denen, die noch gebaut werden zu berechnen, sondern vielmehr über den Grad der Verantwortung der Menschen gegenüber den Errungenschaften der Zivilisation. Die Technologien haben die Eigenschaft, aus der menschlichen Kontrolle zu geraten.

In allen Epochen haben Künstler das Thema der Apokalypse im Fokus der künstlerischen Reflexion. Die Fotografien von Sergey Shestakov sind ein Beitrag, der am Anfang des 21. Jahrhunderts, in der Zeit der Modernisierung, der Innovationen und der neuen Technologien als Vorwarnung interpretiert werden kann.

In seinem Projekt über Tschernobyl ist es Sergey Shestakov gelungen die maximale Artikulierung der künstlerischen Botschaft zu erreichen.

## BIOGRAFIE

## SERGEY SHESTAKOV

Geboren 1968.
Lebt und arbeitet in Moscow.

## STUDIUM

1985-1991
Moscow Physico-Technical Institute

## EINZELAUSSTELLUNGEN

2011
Bhutan. Project “Happiness”. The State Museum of Oriental Art, Moscow
Territory_Pole. Russian State Arctic and Antarctic Museum, St.Petersburg
En marche vers le future. Galerie L’Aleatoire, Paris
Journey into the future. Stop #1. Multimedia Art Museum, Moscow
Extra poles. Mosfilm Gallery, Moscow

## AUSGEWÄHLTE GRUPPENAUSSTELLUNGEN

2011
Antarctic Biennale International Art Exhibition. “Ultima Thule”.
Ca’Foscari University, Venice

2010
The Best of Russia’10. Center of Modern Art “Vinzavod”, Moscow

**SERGEY SHESTAKOV**
**JOURNEY INTO THE FUTURE. STOP #1**

**EXHIBITION / AUSSTELLUNG**

CURATORIAL CONCEPT / KURATORISCHES KONZEPT:
Olga Sviblova

ORGANIZING INSTITUTION / ORGANISIERENDE INSTITUTION:
Multimedia Art Museum, Moscow (MAMM)

MAMM
Multimedia Art Museum, Moscow

**BOOK / BUCH**

DESIGN AND TYPESETTING / DESIGN UND SATZ:
Naroska, www.naroska.de

TYPEFACE / SCHRIFT:
Deck

TRANSLATION RUSSIAN-GERMAN / ÜBERSETZUNG RUSSISCH-DEUTSCH:
Elena Rempel

PUBLISHED BY / VERLEGT VON:
seltmann+söhne, Berlin, Lüdenscheid, www.seltmannundsoehne.de

PHOTO PREPRESS / LITHOGRAPHIE:
Dmitry Shevlyakov

REPRODUCTIONS AND PRINTING / REPRODUKTION UND DRUCK:
Seltmann GmbH, Lüdenscheid, www.seltmann.de

PAPER / PAPIER:
LuxoArt Samt, 170 g/$m^2$

BINDING / BINDUNG:
Buchwerk, Darmstadt



ISBN 978-3-942831-15-4

Printed in Germany

COVER ILLUSTRATION / TITELBILD:
Untitled, from the series
„Journey into the future. Stop #1“

I would like to express my personal thanks to the following people, without whose help neither the exhibitions nor the book could have been realized in their present form: / Ich bedanke mich persönlich bei den folgenden Personen, ohne deren Hilfe weder die Ausstellungen noch das vorliegende Buch hätten realisiert werden können:

Olga Sviblova
Nikolai Kanavin
Patricia Chichmanova
George Chichmanov
Dimitry Shevlyakov
Ekaterina Sholz
Petr Stashuk

Special thank for the support and
patience to my wife Irina Shestakova

Sergey Shestakov